Guy Shrubsole is a writer and environmental campaigner. He has worked for Rewilding Britain and Friends of the Earth, and written widely for publications including the *Guardian* and *New Statesman*. His previous book, *Who Owns England?*, was an instant *Sunday Times* bestseller.

T0047620

'*Who Owns England?* exposed uncomfortable truths about who owns and controls land. His latest offering delves even deeper . . . Shrubsole's campaign should appeal both to those who are nostalgic for times gone by and those more concerned with the future'
New Statesman

'Enchanting and insightful . . . Wonderfully evocative'
Geographical

'Excellent . . . Inspiring'
Unherd

'Utterly enchanting, transporting and spellbinding . . . A rallying cry for restoring the rainforests of Britain urgently, and an inspiring and informative must-read for anyone interested in rewilding and ecological restoration'
Lucy Jones, author of *Losing Eden*

'Passionate, powerful, political and practicable, Guy Shrubsole gives us a blueprint for how to bring our missing rainforests back to life in all their riotous, tangled glory. Impeccably researched, convincingly argued and with generous measures of joyful discovery, this really is a spectacular book'
Lee Schofield, author of *Wild Fell*

'A beautiful, lyrical and urgent book . . . I cannot recommend it enough' **Nick Hayes, author of *The Book of Trespass***

'A treasure chest full of woodland jewels, rare, precious and beautiful' **Chris Packham**

'A magnificent and crucial book that opens our eyes to untold wonders' **George Monbiot**

The Lost Rainforests of Britain

GUY SHRUBSOLE

WILLIAM COLLINS

William Collins
An imprint of HarperCollins *Publishers*
1 London Bridge Street
London SE1 9GF

WilliamCollinsBooks.com

HarperCollins *Publishers*
Macken House, 39/40 Mayor Street Upper,
Dublin 1, D01 C9W8, Ireland

First published in Great Britain in 2022 by William Collins
his William Collins paperback edition published in 2023

4

A catalogue record for this book is available from the British Library

ISBN 978-0-00-852799-0

Set in Adobe Garamond Pro by Palimpsest Book Production Ltd,
Falkirk, Stirlingshire

Printed and bound in the UK using
100% renewable electricity at CPI Group (UK) Ltd

This book is produced from independently certified FSC™ paper
to ensure responsible forest management.

For more information visit: www.harpercollins.co.uk/green

For Mum and Dad;
for my late grandmother, Lilian Jean Maddever;
and for Louisa, with all my love.

CONTENTS

Some temperate rainforests in Britain's 'rainforest zone'

1 Helford River Oakwoods, Cornwall
2 The Dizzard, Cornwall
3 Black Tor Beare, Dartmoor
4 Wistman's Wood, Dartmoor
5 Piles Copse, Dartmoor
6 Lustleigh Cleave, Dartmoor
7 Horner Wood, Exmoor
8 Coed Simdde Lwyd, Ceredigion
9 Llyfnant Valley, Powys
10 Nannau Estate, Gwynedd
11 Coed Crafnant, Gwynedd
12 Ceunant Llennyrch & Coed Felenrhyd, Gwynedd
13 Johnny Wood, Lake District
14 Naddle Forest (Wild Haweswater), Lake District
15 Young Wood, Lake District
16 Ballachuan Hazelwood, Argyll
17 Glen Nant, Argyll
18 Glasdrum, Argyll
19 Lost Valley, Glen Coe, Highlands
20 Ariundle Oakwood, Highlands

SCOTLAND

Edinburgh

IRISH SEA

NORTH SEA

Manchester

Cardiff

WALES

ENGLAND

London

Rainforest zone

And here, as sages say, in days long flown –
Here, on this stormy, barren, blasted ridge,
Luxuriant forests rose; and far away
Swept the bold hills beneath the gazer's eye,
In beautiful succession, dark with leaf –
An ocean of freshing verdure toss'd
By gales Atlantic.

    N. T. Carrington, *Dartmoor*, 1826

1

Paradise Lost

I didn't really believe that Britain was a rainforest nation until I moved to Devon. Visiting woods around the edge of Dartmoor, in lost valleys and steep-sided gorges, I found places exuberant with life. I spotted branches dripping with mosses, festooned with lichens, liverworts and polypody ferns; plants growing on other plants. I was enraptured. Surely, I thought, such lush places belonged in the tropics, not in Britain.

But it's true. Few people realise that Britain harbours fragments of a globally rare habitat: *temperate rainforest*. Rainforests aren't just confined to hot, tropical countries; they also exist in temperate climates. A temperate rainforest is a wood where it's wet and mild enough for plants to grow on other plants. Temperate rainforest is actually *rarer* than the tropical variety: it covers just 1 per cent of the world's surface.[1]

The temperate rainforest 'biome', or set of ecosystems, is strung across the globe, in areas where oceanic currents bring warm winds and torrential downpours. Rainforests exist along the Pacific northwest coast of the USA and Canada; on the southern edge of Chile; in Japan and Korea; across Tasmania and New Zealand; and along the western seaboard of Europe – particularly the Atlantic fringes of the British Isles. In fact, Britain has some of the best climatic conditions for temperate rainforest in Europe.[2]

Awestruck by what I found in Devon, I spent months delving into what's known about these extraordinary places. During my research, I came across an astonishing map made by the ecologist Christopher Ellis showing the 'bioclimatic zone' suitable for temperate rainforest in Britain – that is, the areas where it's warm and damp enough for such a habitat to thrive. This zone covers about 11 million acres of Britain – a staggering 20 per cent of the country.[3] Once, this vast area would likely have been covered with rainforest; but no longer. The *entire* woodland cover of Britain today is just 13 per cent, and much of that is regimented plantations of conifer, planted for timber.[4] Over the millennia, we've destroyed our rainforests, so that now only tiny fragments and relics remain. We're so unfamiliar with these enchanting places, we've forgotten they exist.

As I read more, I started noticing how frequently other habitats get called 'Britain's version of the rainforests'. To the Prince of Wales, peat bogs are 'Britain's tropical rainforests'.[5] To Sir Charles Walker MP, chalk streams are 'England's rainforests'.[6] Whilst peat bogs store billions of tons of carbon and England boasts nearly all of the world's remaining chalk streams, the implication that they're our version of the rainforests is odd.

Britain's *rainforests* are 'our rainforests'.

Yet, as I got used to the idea, it became obvious to me that rainforests belong here. As a country, we're stereotypically obsessed by our rainy weather. How very British, then, to have rainforests. And, as I was to later discover, half-forgotten memories of our rainforests are woven into our myths and legends, and feature fleetingly in poetry and prose from some of our greatest writers.

But why, I began to wonder, have we managed to so comprehensively excise Britain's rainforests from our cultural memory? Why are even environmentalists unaware of their existence? Why was it a surprise to me to find rainforests here, when this should have been something I was taught in school?

Ellis's map of the British rainforest zone describes a damp arc stretching across the west of the country. It's at its most expansive on the wet west coast of Scotland; flows down through the Lake District and parts of the Pennines; washes across Wales' uplands, from Snowdonia to the Brecon Beacons; and soaks parts of the Westcountry, particularly Exmoor, Bodmin Moor and Dartmoor.

I was lucky enough to have moved, by chance, to a part of the Westcountry that abounded in these rainforest fragments – and I quickly fell in love with them. My partner Louisa and I began making frequent visits to them, whenever travel conditions would allow: they became our escape between the coronavirus lockdowns of 2020 and 2021.

It's hard to appreciate the awe and beauty of a temperate rainforest without visiting one yourself. Their rarity and remoteness mean most people in Britain have probably never seen one. So let me take you on a journey into one of Britain's lost, forgotten rainforests.

My first, abiding memory of visiting one of Britain's rainforests is how lush and green it all was. All woods are green in summer, of course: but our rainforests are green all year round, due to the plethora of mosses and lichens clinging to their branches. Even when the leaves have fallen from the trees, they glow with a verdant luminosity. I remember the earthy smell of fungus and leafmould, the distant roar of a river in spate, the drip-drip of falling rain.

A visit to a rainforest feels to me like going into a cathedral. Sunlight streams through the stained-glass windows of translucent leaves, picking out the arches of tree trunks with their haloes of moss. They're places that at once teem with life, and yet have a sepulchral stillness to them. Small wonder that the Celtic druids once revered them as sacred groves.

The trees that make up Britain's rainforests are both familiar and strangely alien. The mainstay of our wet Atlantic woods are oak trees. Yet they're quite unlike the tall, straight oaks of lowland England, so prized by the British Navy in centuries past for supplying it with sturdy timbers for shipbuilding. By contrast, our rainforest oaks tend to be stunted and small, windblown and gnarled – hunched low to the ground to withstand the Atlantic gales, their roots clutching at the thin upland soils.

The limbs of these magical trees are often bent and contorted into outlandish shapes. In their winding trunks, a thousand imaginary faces glower; their twisted branches the stuff of fairy-tales. Out of their boughs have sprung legends of ghostly hounds and horned hunters, Green Men and pixies. Oliver Rackham, one of the greatest historians of our ancient woodlands, referred to them as 'the romantic lichen-hung corkscrew oaks of the deep valleys' of western Britain.[7]

Many of the oaks in our rainforests are sessile oaks, *Quercus petraea*. The Latin name *petraea* means 'of rocky places', which describes well the boulder-strewn hillsides and steep-sided gorges where many of our rainforest fragments still cling on. Even where our rainforests are composed of the more common English oak, *Quercus robur*, few ever reach the height that would make them of interest to foresters. Yet whilst this may have saved them from the axe in times past, during the twentieth century it meant many were cleared to make way for plantation conifers, swapping twisted boughs for telegraph-pole trunks.

Oaks aren't the only species of tree in our temperate rainforests. They also teem with holly, birch, rowan, hazel and ash, the species mix of British rainforests varying in different parts of the country. In windswept upland areas, hardy hawthorn does well, its spiky branches protecting it from munching sheep. On wetter ground, closer to streams and brooks, goat willow thrives. In Scottish rainforests, birch tends to predominate, alongside sessile oaks. The west coast of Scotland also boasts some extremely rare ancient hazel woods, dating back thousands of years.[8]

What really marks out a temperate rainforest, however, isn't the dominant species of tree, but rather the other plants growing on them. Epiphytes – plants that grow on other plants – are a key indicator of rainforest. Epiphytic plants aren't parasites; they simply use the trees as scaffolding, and soak up nutrients from the rain and dampness surrounding them. In Britain, the most common epiphytes are lichens, bryophytes – a grouping that includes both mosses and liverworts – and ferns. The more lichens, mosses and ferns you can see growing on the branches or trunks of trees, the more likely it is that you're standing in a rainforest.

Our temperate rainforests support an abundance of life. They teem with hundreds of species of lichen, many of which are deemed by ecologists to be of international importance. Many rainforest trees are so heavily garlanded with beardy lichens that they resemble Christmas trees hung with tinsel. Others are covered with a spreading filigree of liverworts, running like dark-brown veins over the skin of the bark. On some trunks, the moss grows like deep-pile carpet. Yet others sprout multi-coloured banks of liverworts and lichens which look like coral reefs: purples and greens, silver and gold. The names of these lichens are as rich as their colours: *string-of-sausages*; *tree lungwort*; *witches' whiskers*.

Some branches are so heavily caked in moss that they resemble miniature lost worlds: forgotten plateaus, exuberant with flora. Generations of bryophytes live and die on the same branch, forming a thick, decaying crust like soil, into which the living mosses and ferns can anchor themselves. Life is piled upon life.

Incredibly, these fascinating ecosystems in the air may have a vital role to play in combating the climate crisis. All forests remove carbon from the atmosphere, locking it up in tree trunks and in the soils of woodland floors. But rainforests may also be locking up carbon in the soil that forms *in their canopies*. Researchers examining the tropical rainforests of Costa Rica have recently found that the epiphytic plants and decomposing leaf-litter that accumulate on the branches of rainforest trees can store more carbon than the soils on the ground. 'It could be a substantial carbon sink, and we need to account for it,' they argue.[9] Could Britain's temperate rainforests, their branches piled high with mossy humus, also be performing this crucial function?

Lichens and mosses are small and can be hard to identify; but easier to spot from a distance are the polypody ferns that sprout from Britain's rainforest trees. These beautiful, delicate plants love the damp shade and humidity of our western oakwoods. The translucent fronds of polypodies lend a furry green aura to each tree they grow on, particularly when sunlight glows through them. 'Polypody', meaning 'many-footed', refers to the fern's method of spreading. It sends out horizontal roots called rhizomes that creep along branches, step-by-step, sending up fresh shoots at each interval.

Occasionally you find other fern species growing epiphytically on trees, too. In some of our dampest woods, I've seen huge clumps of bracken protruding from standing trunks, where a spore has lodged itself in a knot-hole. Surrounded by these seemingly primeval plants, I'm transported backwards in time. Back before flowering plants first appeared; back to the time of the dinosaurs, over 65 million years ago, when ferns first evolved. Rainforests, then, remind us of when giants walked the Earth, when brachiosaurs would have lumbered through the swamps, eating the fronds of ferns amidst the spore-laden air.

These ferns also transport me across the world, to where huge tree-ferns grow in tropical rainforests from Brazil to Indonesia, as well as in the southern temperate rainforests of New Zealand. To Oliver Rackham, these British woods, their branches festooned with epiphytes, were 'the last northern outlier of the massive "fern-gardens" high in the trees of tropical rainforests'.[10]

These miniature ecosystems are home to some startling phenomena. On one visit to a patch of Devon rainforest, I was astonished to spot a tree growing in another tree. A holly sapling had sprouted from the moss-encrusted bole of a beech, rearing up impudently at a height of some ten feet above the forest floor. *How on earth did that get there?* I wondered. Presumably a bird had dropped a holly berry – either out of its beak or its bottom – and, finding the mossy beech to be fertile ground, it had put down roots.

The writer and RSPB site manager Lee Schofield calls these minor miracles 'air trees'. Lee manages a nature reserve deep in the temperate rainforest zone of the Lake District. Comparing our finds one day, he showed me a photo he'd taken of a hawthorn, a rowan and an oak sapling, all growing together in the arms of a mighty ash tree. Somehow, nature had conspired to produce this aerial ecosystem, a kind of miniature kingdom in the clouds, sustained by the extreme dampness of the local climate.

But rainforests are home to more than just trees. Their canopies support a rich and diverse bird fauna. Besides the usual species of British woodland birds, our western oakwoods are particularly good habitats for threatened wood warblers, declining pied flycatchers, and the fire-tailed redstarts.[11] I have watched treecreepers scamper up the trunks of rainforest trees on Dartmoor, and listened to the yaffling cry of green woodpeckers as they search for grubs and insects in the bark. Overhead, buzzards mewl, and the air occasionally croaks with the guttural *ak-ak* of a passing raven. One bird in particular is vital to the survival of Britain's rainforests: the jay, with its blue-and-black striped wing feathers. Its appetite for acorns makes it nature's chief planter of oak trees. Jays will carry acorns far from their mother tree, bury them in the earth for safekeeping and then forget where some of them have been stashed; the survivors grow into new oaks.

Tear your eyes away from the riotous life of the overstorey, and the understorey of our rainforests is no less exciting. The polypodies that throng the branches are joined on the forest floor by a wider array of fern species, both hard ferns (*Blechnum*) and soft (such as *Dryopteris*). Unfurling fronds resemble the curled scroll of a violin. Hart's-tongue fern lolls from earthen banks, fleshy and glistening. Bilberry (*Vaccinium*), a low-growing shrub often found on moorland, also grows well in Atlantic woods, offering up its blue fruit to grazing animals and foragers. Hedgerows and holloways in temperate rainforests are cascades of plants: pennywort, with its crisp, circular leaves; bitter-smelling woundwort; effervescent meadowsweet. Cataracts of bracken gush out of the damp, moss-coated banks, studded with jets of red campion, brambles and sticky-bud.

In spring, many of our rainforests become bluebell woods. The poet Gerard Manley Hopkins wrote of how the 'azuring-over greybell makes / Wood banks and brakes wash wet like lakes'.[12] The wet Atlantic woodlands, already awash with moisture after winter rains, transform into seas of flowers that shimmer with a near-phosphorescent blue.

Perhaps so far I've given too sunny a description of these rainy forests. Their usual weather, of course, is wet. *Really wet.* If you're lucky enough to visit them when the sun's shining, you're not really seeing them at their best. For most of the year they drip continuously with moisture, drip from their canopies, drip from every branch and leaf and stem. The rain runs through these woods in streams, along footpaths that *become* streams in winter torrents. You'll return from a visit to a British rainforest soaked from head to toe, but feeling all the more alive for it.

Whilst rain brings life, it also brings decay: dead wood rots. But that decay in itself brings forth more life; fungal life. Gelatinous wood-ear fungus spawns from decomposing branches. Giant bracket fungi protrude from rotting trunks, large as dinner plates; their upsides hard like clam shells, their undersides white and fleshy. Most outlandish of all is the hazel gloves fungus, unique to temperate rainforests, whose orange lobes resemble creeping fingers.

Frequently, the fungal kingdom appears to overlap with the kingdom of faerie. Toadstools form fairy rings in glades, the product of a fungus's mycelial roots spreading outwards in a circle before fruiting. Once, on a visit to a secluded rainforest valley, my eye was caught by a flash of red amidst the overwhelming green of the wood. A tiny scarlet elf cup was growing amongst the ferns, so-called because, in folklore, elves were said to drink from them.

Though they're harder to spot, the understorey is also home to many species of mammals, reptiles and insects. Mice and shrews scuttle through the thickets of scrub; adders lurk in the grass; silvery badgers and lean foxes betray their silent presence with a trail of droppings. Roe deer (and, in Scotland and on Exmoor, red deer) browse on the lower branches of the trees. Grey squirrels have now almost entirely replaced native red squirrels in England's woods, but they cling on in Welsh and Scottish rainforests. Pine martens and stoats prey on rabbits and rodents. Invertebrates abound in the rotting wood and rich leafmould of our Atlantic woods. Amongst many other species, Dartmoor's rainforests are home to the exceptionally rare blue ground beetle, which has been spotted in only a handful of locations in the country.

Many of the larger mammals that once populated our temperate rainforests are now sadly long gone – elk, boar, lynx, wolves – or on the verge of extinction, like wildcats. But beavers, wiped out 400 years ago by the fur trade, are now starting to be reintroduced to our western woods and rivers as more people recognise their vital importance to the ecosystem. Other species may yet follow.

Britain's rainforests, in short, are truly the pinnacle of our country's woodlands. Not only are they extraordinary places to experience, providing a feast for the senses. They're also treasure-troves of biodiversity, home to globally significant populations of rare species of lichens and mosses, birds and mammals. And the carbon that our rainforest trees are busily soaking up – not just in their trunks, but also via the epiphytic plants that festoon their branches – make them some of our best allies in the fight against the climate crisis.

Britain's rainforests conceal other secrets, too. On some of our adventures, my partner and I would find the overgrown ruins of forgotten buildings: abandoned farmsteads and tin-miners' huts, their tumbledown walls carpeted in moss and enmeshed in tendrils of ivy, like lost Mayan temples.

Who once lived here? Why were these places abandoned?

These mysteries would later lead me deeper into the histories of our rainforests. And contained in those histories of human exploitation and interaction, I started to find seeds of a possible future; of how Britain's rainforests might one day return and spread.

When I was five years old, my Mum organised a 'Save the Rainforests' party in our back garden to raise money for Friends of the Earth. We painted a mural for our local library: colourful toucans and parrots and rainforest trees hung with vines. The year was 1990, and the cause célèbre of western environmentalism was the battle to save the Amazon.

The word 'rainforest' was first coined by an obscure German botanist in 1898, but only came into common currency in the late twentieth century. Previously, tropical rainforests had been known to westerners simply as 'jungles' – originally a Sanskrit word meaning 'rough and arid', but freighted with colonial connotations. 'Jungle' carries with it Victorian ideas of 'darkest Africa' and the Conradian heart of darkness; of imperial explorers battling 'savage' tribes; of Tarzan, Lord of the Jungle; of Kipling's *Jungle Book* and 'the law of the jungle', in which only might is right. When the British geographer Lieutenant-Colonel Percy Fawcett set out on his doomed expedition to find the fabled Lost City of Z, he ventured into the Amazon *jungle*, not the Amazon rainforest. Jungles were hostile places to be exploited and overcome, and ideally hacked down with machetes. They certainly were not places that needed preserving.

But after the post-war economic boom led to soaring global demand for Brazilian rubber and Indonesian palm oil, the machetes became axes and the axes became chainsaws. The jungles, previously so vast, began to shrink before the onslaught of development. Since the last Ice Age, humanity has cut down a third of the world's forests; half of this destruction has happened in the last century alone.[13] The rich world's hunger for timber, for polished mahogany tables and teak balustrades, was matched only by its lust for Amazonian gold, its thirst for coffee and tea and its consumption of rainforest beef. Tropical deforestation, which did not begin in earnest until after 1950, reached rates of 12 million hectares a year by the 1990s: an area of rainforest around half the size of the UK, lost annually.[14]

This same orgy of destruction generated its own antithesis in the form of the modern environmental movement. The first images of the Earth from space, beamed back by the Apollo astronauts in the late 1960s, helped give a sense of the planet's fragility and demonstrated, in starkly visual terms, the limits to economic growth. Between the first Earth Day of 1970 and the Rio Earth Summit of 1992, millions of people across the world rose up to demand action against air and water pollution, environmental destruction and the despoliation of the natural world.[15]

Though this was the era of Save the Whales and of Live Aid, when it felt like rock music and ethical consumerism might just save the world, the early environmental movement was not solely western in nature. In fact, the first modern 'tree-huggers' were the Chipko women of Uttar Pradesh in India, who chained themselves to trees to stop them being cut down. But environmentalism in the 1970s and 80s did find a particularly receptive audience amongst the western middle classes. And it was into this context that the movement to save the tropical rainforests was born.

First, though, western audiences needed to know what it was they were saving. Environmental campaigners alighted upon the word 'rainforest' as an alternative to 'jungle', free of the old term's imperial baggage. 'Rainforest' was more accurately descriptive of the habitat, for one thing; equally important, it hinted at something beautiful, fragile and in need of protecting. 'A primary strength of the Save the Rainforest movement was that it was focused on something positive and amazing. The rainforest was magic, and saving it was fun, not grim,' writes the journalist Dan Nosowitz.[16]

Interestingly, one of the first rainforests to come to the attention of the western world was a temperate rainforest, not a tropical one. The fight to stop the Franklin Dam in Tasmania, and save an entire river valley carpeted with old-growth rainforest, convulsed Australian politics in the late 1970s and early 80s. On one side stood the Tasmanian Hydro-Electric Commission, hell-bent on damming the Franklin, 'Australia's last wild river', and drowning the pristine forest. On the other side was an army which rose up in defence of Tasmania's rainforests, spanning conservationsts and local residents, hippies and accountants. 'Most were people who had never rafted those wild rivers, or walked in the majestic silence of the threatened forests . . . But they were people with a fierce love of nature, who in wilderness saw nature at its purest,' wrote one of the activists.[17] Thousands of them would go on to commit criminal trespass by blockading the river with kayaks, preventing bulldozers from operating, and unfurling banners quoting Thoreau ('In wilderness is the preservation of the world'). It worked; and in the ensuing political firestorm, the blockaders brought down the Australian government. The incoming Labour Prime Minister Bob Hawke's first words upon taking office were: 'The dam will not be built.'[18]

But whilst one fragment of Tasmania's rainforest had been saved, wholesale destruction of other rainforests continued apace across the globe. From the mid-1980s onwards, the movement to save the rainforests was dominated by the plight of the Amazon. Brazil's military junta had set their minds to developing mighty Amazonia, driving the Trans-Amazonian Highway through its heart and opening it up to loggers, miners and cattle ranchers. In the UK, the first Save the Rainforests campaign was started by the environmental charity Friends of the Earth in 1985 – the same campaign that my mum would go on to raise funds for and that I would paint parrots in aid of. That same year saw the founding of Rainforest Action Network in the USA, a new direct action group that hung banners off buildings and picked fights with corporate despoilers of the Amazon.

Alongside the direct action and political mobilising was a hefty dollop of green consumerism and cultural activism. In the wake of Live Aid, there were benefit concerts for the rain-forest; Sting set up a foundation to help buy rainforest land for preservation. Friends of the Earth warned consumers off buying tropical hardwoods with the slogan 'Mahogany is Murder', and pointed them towards sustainable suppliers in their *Good Wood Guide*. Burger King and McDonald's faced consumer campaigns railing against their reliance on rainforest beef. The Body Shop started selling Brazil-nut-based shampoos; Ben and Jerry's devel-oped 'Rainforest Crunch' ice cream; and the Rainforest Alliance started certifying coffee and bananas that had been grown without relying on tropical deforestation.

It would be easy to mock these efforts as tokenism by western consumers trying to assuage their guilt. Yet all these forms of protest were as vital as they were of their time. They raised awareness of the plight of the rainforests, whilst helping stem a little of the west's voracious demand for consumer products that relied on tropical deforestation.

Ultimately, of course, it took deeper shifts in Brazilian domestic politics to really make a dent in stopping the destruc-tion of the Amazon: the end of the junta, the death of the rubber-tapper turned activist Chico Mendes, the recognition of Indigenous rights in a new democratic constitution and the election of environmentalist President Lula da Silva in 2003. Having reached a terrifying peak in the early 2000s, Amazonian deforestation nosedived by 80 per cent between 2004 and 2012 because of two key factors: stronger government enforcement of rainforest protections, and international consumer pressure on companies to place a moratorium on sourcing soybeans from the deforested Amazon.[19]

As a clarion call and a movement, 'Save the Rainforests' was a formative moment for modern western environmentalism. I have it to thank for my own career choices, too: my early foray into banner-painting helped lead me to later join Friends of the Earth as a campaigner. But something the Save the Rainforests movement didn't teach me as I was growing up was that Britain, too, possessed rainforests. It made the problem appear to be 'over there', rather than something taking place in our backyard. Ironically, whilst we were all trying to save the rainforests of the Amazon, we didn't realise that we'd destroyed our own back home.

'Today's European rainforests are mere fragments,' writes ecologist Dominick DellaSala: 'A reminder of a bygone era when rainforests flourished, they are barely hanging on as contemporary rainforest relicts.'[20] But it was not always so.

Once upon a time, long, long ago, Britain was covered in many more trees than it is today. Oliver Rackham called this vast expanse 'the Wildwood'. Between the end of the last Ice Age around 12,000 years ago, and the start of the Neolithic period some 6,000 years ago, tree cover advanced until 'almost the whole of Britain was covered with virgin forest which was just beginning to be affected by human activities'.[21] The retreat of the glaciers and the advent of a warmer, wetter climate allowed trees to spread north into Britain, colonising the lands scraped bare by the ice. As the ice melted, sea levels rose, forming the English Channel and cutting Britain off from continental Europe.

The Wildwood developed in character over thousands of years, with successive waves of colonisation by different species, as climatic conditions shifted and changed. Ancient pollen records suggest that birch came first, advancing over the post-glacial tundra, followed by pine. Then came hazel, oak and elm, and later lime. The mix of trees increased in diversity, as well as in area.[22] One of the first historians of Britain's ancient flora, Sir Harry Godwin, wrote that whilst the retreat of the ice at first 'favoured species of open habitats', later periods witnessed a 'thickening sea of woodland'.[23]

In recent years, the idea that Britain's prehistoric tree cover comprised continuous woodland has been challenged by the ecologist Frans Vera. Instead, he argues, it seems more likely that the primeval Wildwood was a shifting mosaic of closed-canopy woods, open glades, scrub and grasslands. This mix of habitats moved in dynamic cycles because of grazing and disturbance by wild animals, many of them now extinct. Wild aurochs, for example – similar to bison and some native breeds of cattle – would have knocked over trees, browsed branches and eaten saplings. Beavers, once widespread in Britain and only now being reintroduced, would have created clearings in the forest by gnawing down trees for their dams.

Vera's thesis – which was largely accepted by Rackham at the end of his life – points to Britain's prehistoric landscape being more open in character than the dense, primeval forest of popular imagination. Yet even though a squirrel might not have been able to leap from tree to tree from Land's End to John o'Groats, as the old adage has it, it's clear that Britain still had many more trees and woods then than it does today.[24]

Britain certainly had more rainforests then, too. By the time the Wildwood had fully developed (around 4000 BC), woods of oak and hazel had become the dominant woodland type in the west of Britain, stretching from the western Highlands of Scotland down through the Lake District, the Pennines, Wales and southwest England. This 'oakwood province' corresponds closely to the temperate rainforest zone identified by climatologists.[25] The mild and rainy climate of this time – the start of the New Stone Age – is known to archaeologists and geologists as the Atlantic Period. It's most likely the phase during which our temperate rainforests were at their greatest extent, covering perhaps one-fifth of Britain.[26]

Dartmoor, near where I live, is mostly treeless today – but the peat bogs that cover it contain much buried evidence of lost rainforests. Because they're an anaerobic environment, peat bogs are capable of preserving organic matter for thousands of years. Drilling down into this peat to produce 'peat cores', filled with ancient pollen samples, offers up a stratified history of past habitats. The geographer Ian Simmons, who studied Dartmoor's prehistoric pollen records in the 1960s, found the peat to be full of pollens produced by birch, pine, hazel, willow, holly and oak. Simmons concluded that 'on climatic grounds, there seems no reason why the whole upland should not have been forested during the "forest maximum"'. Crucially, he also found remains of spores from the polypody fern throughout this same period: evidence of epiphytes growing on trees in this ancient rainforest hotspot, some 5,000 years ago.[27]

Fossilised tree remains corroborate the picture that Dartmoor was once heavily wooded. A study by two geographers, David Maguire and Chris Caseldine, mapped all the discoveries of 'bog oaks' unearthed in an area of northern Dartmoor – an area today completely denuded of trees. They found a particular distribution of former oakwoods along river valleys, but also declared: 'It now seems likely that forest covered nearly all of Dartmoor' up until around 5,000 years ago, 'although there remains a possibility that the highest elevations carried moorland.'[28] Reviewing studies of other upland areas of Britain, they found the historical 'tree line' – the point on a mountain above which no trees grow naturally – to have been far higher than we now assume. In the Lake District, for example, the tree line during the Atlantic period is estimated to have been around 760 metres above sea level.[29] Today, the highest wood in the Lakes – Young Wood at Bowscale Fell, a fragment of temperate rainforest – lies at an altitude of just 485 metres.[30]

So what happened to our rainforests, and the Wildwood as a whole?

The awful truth is that we destroyed them. As Stone Age hunter-gatherers developed agriculture and started settling, they began to clear the Wildwood for farmland with flint axes, fire and browsing livestock. It was undoubtedly hard work: native broadleaved woodland 'burns like wet asbestos', in Rackham's pithy phrase, making it hard to simply set light to the forests. But the process of clearance was probably aided by burning tree stumps after felling, and putting cattle and sheep to work browsing the regrowth shoots, preventing the trees from regenerating. In other words, many of Britain's rainforests fell to the same sort of slash-and-burn agriculture practised today in the Amazon.

As stone tools were replaced first by bronze ones and later by iron, hacking down the Wildwood became easier. The zone of destruction also moved from the lowlands to the uplands. 'During the Bronze Age (1700–500 BC)', Rackham wrote, 'clearance of the Wildwood continued and extended into high altitudes.'[31] Dartmoor, for example, bears the remains of many Bronze Age hut circles and settlements. To Ian Simmons, 'it is inconceivable that, coming into an upland whose flanks and valleys were still forested . . . these people did not effect a considerable disforestation'.

Yet many areas of rainforest escaped the destruction of the Wildwood, surviving into more recent centuries. As later chapters will show, some lasted long enough to make their way into the historical record, their after-echoes etched into place names, myths, legends: Wistman's Wood on Dartmoor, Coed Felenrhyd in north Wales, Keskadale in the Lakes. Some succumbed to medieval tin-miners; some to Victorian charcoal-burners; others to the hungry, nibbling teeth of sheep. Unforgivably, some of Britain's rainforests were destroyed as recently as the twentieth century, felled by landowners and the Forestry Commission in the name of timber production.

Britain was once a rainforest nation. But we lost most of our rainforests.

The first map of temperate rainforests worldwide was published shortly after the Rio Earth Summit in 1992, when world leaders gathered for the first time to address climate change and bio-diversity loss. Using what were then state-of-the-art digital methods – known as geographic information system (or GIS) mapping – the study, by an Oregon-based NGO called EcoTrust, revealed the rarity of a habitat that covers less than 1 per cent of the globe.

'Like the tropical rain forests which have rightly received so much attention, these forests are an important part of our global heritage,' wrote the report authors. The resulting map appears pixellated and crude to modern eyes, offering few details for anyone wishing to zoom in on Britain, say, to find their local rainforest. But the study was clear about how little temperate rainforest was left nowadays: 'Almost total manipulation and transformation of natural vegetation cover in Europe has done much to undermine a cultural sense of the presence of temperate rain forests in the region.'[32]

This loss of cultural memory, this great forgetting that we once had rainforests, is almost as heartbreaking as the loss of the forests themselves. It points to the phenomenon that ecologists call 'shifting baseline syndrome': society's ability to grow accustomed to environmental losses. What appears to us today as a 'green and pleasant land' is, in reality, a desert compared to the glory of what once existed. *There were giants on the Earth in those days*. If we are to stand any chance of restoring our lost rainforests, we first need to remember we once had them.

Not everyone *has* forgotten, of course. Part of the job of environmentalists is to see the damage wrought by humanity, no matter how much it might hurt to do so; to see a 'world of wounds', as the early conservationist Aldo Leopold put it.[33] Many ecologists have already begun the job of diagnosis. I first learned about the existence of temperate rainforests from the environmental activist George Monbiot's book *Feral*, in which he briefly recounts a visit to a fragment of Atlantic rainforest in mid-Wales.[34] The conservationist Clifton Bain has visited and photographed many of Britain's surviving rainforests, recounting them in his travelogue *The Rainforests of Britain and Ireland*.[35] Dozens of pioneering botanists, bryologists, lichen specialists and woodland ecologists have explored these habitats over the decades, publishing academic papers and raising the alarm about the ongoing threats they face. Their voices feature throughout this book.

And yet . . . Britain's rainforests remain under-recognised, unmapped, and largely unacknowledged by politicians and a wider public. 'Because very few rainforests remain throughout Europe, they have not received much attention from ecologists,' notes Dominick DellaSala, one of the few authorities on this habitat.[36] I read every study I could find voraciously, following in the footsteps of those who've explored our rainforests before. But as I went further, the footsteps became fainter to see, the path overgrown. Why was there no proper map of where Britain's temperate rainforests survive, I wondered? The more experts I spoke to – the more lichenologists, climatologists and mapmakers I called on for help – the less certain I became that this precious habitat was fully understood.

So I decided to start a blog, Lost Rainforests of Britain, to try to drum up public support for protecting and restoring these amazing places. I created a Google Map to gather together examples of where fragments of our rainforests cling on, adding locations that Louisa and I had visited.[37] To launch the blog, I put a call-out on Twitter for people to send in photos of potential rainforest sites they'd visited, thinking that at best it might generate a handful of new leads.

I was overwhelmed by the response. Hundreds of people sent me submissions for the map, deluging me with photos of the beautiful wet woodlands that they knew and loved; photos of trees covered in mosses and ferns, thriving in hidden valleys and inaccessible gorges up and down the western edge of Britain. My email inbox overflowed with excited messages of support, expressing surprise and delight at the realisation that the woods they'd been walking in turned out to be temperate rainforest. A follow-up piece I wrote for the *Guardian* received over 200,000 views.[38]

It felt like I had struck a chord. Perhaps it was just curiosity at the fact that we have rainforests here in Britain; perhaps it was part of the wider public yearning to reconnect to nature, brought about by months locked down in our homes. But maybe this subject also taps into something deeper in the national psyche: the sense that, through our destructive actions, we have lost something profoundly important from the natural world.

We are haunted by a folk memory of the great Wildwood that once covered Britain, whose outlines occasionally resurface not just in our myths but also in pollen cores and fossil evidence. The biblical story of the Fall – that we once lived in paradise, but lost it due to our sins – remains a powerful narrative, even in a post-Christian country. The Save the Rainforests movement of the 1980s and 90s brought home to us the terrible destruction still being wrought to the earthly paradise of the tropical rainforests. Doing something to repair the damage done to our own rainforests here in Britain – 'this other Eden', this 'green and pleasant land', as our greatest playwrights and poets would have it – feels like a redemptive quest. In restoring them, we might yet restore a missing part of ourselves.

Because our rainforests aren't irrevocably lost. Fragments survive. And as I was to discover, in some parts of Britain rainforests still thrive. Far from being dying relics from some bygone era, they're living ecosystems – growing, regenerating and spreading, whenever they're given half a chance. As Jeff Goldblum memorably intones in *Jurassic Park*: 'Life . . . finds a way.'

This book is the story of my quest to find Britain's lost rainforests, and bring them back.

2

Ghosts in the Landscape

The 2nd of April 1901. Sir Arthur Conan Doyle, the great crime novelist, has just finished a plate of scrambled eggs, sausages and bacon in the breakfast-room of the Old Duchy Hotel on Dartmoor. Dabbing the crumbs from his thick handlebar moustache, he lays the napkin on top of the discarded bacon rinds, and reaches for his tobacco pouch and pipe. A match flares, and is then extinguished. And then, taking out his fountain pen, he begins to write:

> *Robinson and I are exploring the moor together . . . It is a great place, very sad and wild, dotted with the dwellings of prehistoric man, strange monoliths and huts and graves . . .*[1]

Conan Doyle writes to his mother like this every few days. He has journeyed to Dartmoor to meet his friend, the journalist Bertram Fletcher Robinson, who has lured him there with tales of the supernatural; tales that have inspired Conan Doyle to begin writing a new Sherlock Holmes novel. Robinson will later recall how the author 'listened eagerly to my stories of the ghost hounds, of the headless riders and of the devils that lurk in the hollows – legends upon which I had been reared, for my home lay on the borders of the moor'.[2]

A fire cracks and hisses in the hearth, its flames feeding not on coal but on peat dug from Dartmoor's blanket bogs. The latch on the great oaken front door to the hotel lifts with a clatter, letting in the gale blowing outside. Conan Doyle looks up to greet Robinson, who has just come in from the stableyard, rain dripping from his coat. Another man is with him: shorter and stouter than Robinson, his skin weather-beaten and ruddy from a life lived outdoors.

'Arthur, I'd like you to meet our coachman,' says Robinson. 'He'll be showing us around the moor today. His name is Harry Baskerville.'

We can't be sure of the route Conan Doyle and Robinson took that day over Dartmoor. But we do know from the novelist's letters that they covered around fourteen miles. And we know several of the people they met and places they visited, because they found their way into Conan Doyle's subsequent thriller, called – of course – *The Hound of the Baskervilles*.

Harry Baskerville really was the name of the author's coachman on his trip. Conan Doyle also drew inspiration for the character of the wicked Hugo Baskerville from tales of a local seventeenth-century squire, Richard Cabell, whose reputation for evil was such that he was buried under a heavy stone behind iron railings, to prevent his spirit rising again and doing the Devil's bidding.

The Grimpen Mire, where Conan Doyle's spectral hound is kennelled, is partly based on Grimspound – a Bronze Age settlement on the moor's eastern flank – and partly on Foxtor Mire, a morass just to the south of Princetown, where the novelist was staying. It would have been a short hike for Conan Doyle to stride out to the Mire and survey its vast emptiness, a great expanse of blanket bog, treacherous to unsuspecting travellers.[3]

What of the legend of the hound itself? Dartmoor has no shortage of shaggy dog stories. The wicked Richard Cabell's grave in Buckfastleigh was said to have been visited by a pack of baying phantom hounds on the night of his burial. But the tale of ghostly hounds roaming Dartmoor comes from another location close to where Conan Doyle was staying, one that has its own long history and deep associations with myth and folklore: Wistman's Wood.

Wistman's Wood is no ordinary wood. It's an ancient fragment of temperate rainforest, suspended on a hillside on the upper reaches of the West Dart River. It first crops up in the historical record in the early 1600s, around the time of Shakespeare, and tree-ring studies suggest individual trees could be hundreds of years old.[4] But the wood may well have stood here for *thousands* of years, stretching back into prehistory, a remnant of primeval forest. Wistman's Wood is awe-inspiring: a tangle of stooped and shrunken trees, their twisted branches festooned with mosses and lichens. Oaks and occasional rowans emerge from an avalanche of moss-covered monoliths, a stone carapace protecting the younger trees from the depredations of nibbling sheep.

The inner recesses of Wistman's Wood are almost impassable, both to sheep and human visitors. The trees cling on to this boulder field for dear life, their crowns flattened horizontal against the weather. Dartmoor's howling winds have shaped them like a demonic topiarist, stunting each ancient oak into a bonsai tree. Grimacing faces appear in the gnarled and knotted trunks, bringing to mind the Ents of Tolkien's Fangorn Forest. Wreathed in autumnal mists, the wood feels Otherworldly, and with good reason. It is as soaked in legend as it is in rain.

The very name 'Wistman's' derives from the Devonshire dialect word *wisht*, meaning 'eerie, uncanny' or, in some readings, 'pixie-haunted'.[5] For centuries, the wood has been renowned amongst visitors as one of the wonders of Devon. Ghosts and horror stories abounded. Wistman's Wood fired the Victorian imagination, speaking to its obsession with the

Gothic, the ancient, the primeval that lies beneath the veneer of modern civilisation. Folklore hunters whispered apocryphal tales about it being a sacred oaken grove, in which Celtic druids once committed blood-sacrifice to ancient gods. It was seen as a domain of pixies and fairy folk, a 'thin place' where the gap between our world and a spectral Otherworld is narrow and passable.

Go to Wistman's Wood on a wintry day in fog, and you can well believe it. I once visited the wood on New Year's Eve, when it was wreathed in mist, each moss-encrusted branch glistening with dew. The air between the trees was translucent, causing each coiling limb to twist and coalesce into macabre shapes. An eerie silence pervaded the wood, and the enveloping fog seemed to cast it adrift from the surrounding moor, as if it were lost at sea. Walking alone, with twilight falling, I suddenly stopped in alarm. There, crouched between the gnarled oaks, was the unmistakeable shape of a black hound. I stood frozen to the spot, my heart pounding. Slowly, my conscious brain regained control from my overactive imagination: it was just a fallen branch trapped between moss-covered boulders. But it took a while for my pulse to return to normal.

Wistman's Wood is indelibly associated with the folk tale of the Wild Hunt: a supernatural chase in which spectral huntsmen ride out across the moor, led variously by the Viking god Odin, the Fairy King or the Devil himself, depending upon whose account you read. Joining the phantom riders on their hunt are jet-black dogs known as 'wisht hounds'. It is these, surely, which inspired Conan Doyle's diabolical hound.

We don't know if Conan Doyle visited Wistman's Wood during his stay in Princetown, but given its proximity and allure, and his fascination with the supernatural, we might assume he did. After all, *The Hound of the Baskervilles* makes several references to the shrunken oakwoods of the moor, nestled in their rocky hollows. At one point, on the approach to Baskerville

Hall, Dr Watson refers to the 'stunted oaks and firs which had been twisted and bent by the fury of years of storm'. Elsewhere, he describes what we now recognise as the luxuriant temperate rainforest climate of Dartmoor, travelling through a deep holloway 'heavy with dripping moss and fleshy hart's-tongue ferns . . . Both road and stream wound up through a valley dense with scrub oak.'

This landscape forms the backdrop to the unfolding drama of the novel. Yet it is also at its heart. As the novelist John Fowles wrote in an essay about Wistman's Wood, 'the real black hound is the Moor itself – that is, untamed nature, the inhuman hostility at the heart of such landscapes'.[6]

But there is also something else haunting about Wistman's Wood and the barren moor that surrounds it – a nightmare that stalks the twenty-first century imagination far more than tales of ghost hounds and bogeymen. It is the spectre of ecological collapse. Forget the ghost stories: the real ghosts in Dartmoor's landscape are the ones rising from the bones of the rainforests that we destroyed.

The ecologist Ian Rotherham has coined the term 'ghost woods' to describe landscapes emptied of trees. As we now know, after the last Ice Age, Britain was far more wooded than the pitiful 13 per cent woodland cover we have today.

In some parts of the country, the spectral presence of these ghost woods are still discernible by 'indicator species' of certain plants, like bluebells and bracken. Some ecologists speculate that the bracken fringe surrounding Wistman's Wood may indicate a former woodland soil, stretching far beyond the present tree line.[7] There's an old saying amongst hill farmers – 'under bracken there's gold' – which speaks to the richer, deeper soils that lie beneath bracken stands and may be a remnant of former forest floors.[8] Wood pasture can be another good indicator. This savannah-like landscape, where occasional lone trees survive amidst grasslands grazed by sheep and cattle, can point to what was once denser woodland. It's even possible to make out the ghosts of lost woods in old names and maps.

But Rotherham's search for ghost woods isn't intended merely to mourn what's departed, rather to resurrect what still survives. As he writes, 'The ghosts are there in moors, heaths and grass-lands, from extensive grazed "woods" to tiny pockets of roadside verge. You only have to look.'[9]

Reading this, I decided to go looking for ghosts on Dartmoor. If I could find the remnants of ghost woods, I reasoned, maybe I could piece together a picture of how the moor used to look in the dim and distant past. Perhaps by unearthing the remains of long-gone rainforests, I could help resurrect them in the future. Surveying the treeless landscape of Dartmoor today, you might think such a quest would be in vain. Yet when you start looking, trees and woods sprout from Dartmoor place names.

Some of Dartmoor's ghost woods are obvious: Birch Tor, Thornworthy Down, Hawthorn Clitter (the latter referring to a boulder field). Others are more hidden, concealed in the landscape like a Dartmoor Letterbox. The River Okement, which flows through Okehampton – a town on the moor's northern edge – likely gets its name from the oak trees that still grow along its edges. Not far from Venford reservoir flows a mono-syllabically named stream, the O Brook. The writer Adrian Colston theorises that the 'O' is an abbreviation of 'oak', and that 'we might speculate that before the valley came to the attention of Mediaeval tinners it would have been another high level oak wood like Wistman's Wood'.[10]

'Watern Oke' appears cryptically on modern Ordnance Survey maps of Dartmoor, sandwiched between the River Tavy, some Gothic-scripted hut circles, and an MOD firing range. But look back at the first edition OS maps from 1809 and you find 'Watern Oak', next to a tiny engraving of a lone tree. An eyewitness account by a visiting rambler in 1867 described the

Watern Oak as a 'solitary tree, backed by the abrupt crags and lofty peak' of Fur Tor.[11] A Victorian antiquarian, John Page, wrote in 1895 of 'the solitary wind-twisted tree called Watern Oak, which has been so often hailed with delight by the moor-man lost in a fog'.[12] Alas, no such tree exists today.

Follow the Tavy further upstream to Fur Tor, and you come across another wood that isn't there. It's not marked on modern maps, but William Crossing's classic *Guide to Dartmoor* (1909) records 'a spot known to the moormen as Fur Tor Wood. The name seems to point to the former existence of trees in this sheltered hollow, and the discovery a few years ago of oak buried in the peat near Little Kneeset proves that they once grew around here.'[13] It's likely that the boulder-strewn sides to Fur Tor once provided cover for tree saplings to thrive, in the same way that they still do at Wistman's Wood and elsewhere.

There's disagreement about how Fur Tor, or Vur Tor, got its curious name.[14] One local historian, Alexa Mason, has suggested to me that it could be derived from 'Furze Tor', the old name for gorse. But I wonder if it's a corruption of *vert*, the French for 'green', hinting at its formerly wooded nature. The old system of 'venville rights' which once governed Dartmoor permitted commoners to take 'all thinges that maye doe them good, savinge vert (which they take to be green oke) and venison'.[15] So deer and oak trees were off-limits to commoners hungry for meat and firewood. Perhaps Vert Tor was a warning sign: *get orf my land*. Clearly, it didn't work. Fur Tor Wood is no more.

Dartmoor has a wealth of weird and wonderfully named tors, and perhaps my favourite is called 'Dunna Goat'. But what sounds like the confessions of a sheepish rambler is, in fact, a clue to a forgotten woodland. Various sources agree that the name of this rocky tor overlooking the Rattle Brook on the eastern side of the moor is actually a corruption of the Celtic *dun-y-coed*: the 'wooded hill'. Whatever once existed here must have been cut down or grazed away long ago. Today, it is a howling waste of purple moor-grass tussocks, interrupted only by the ruins of a peat-cutter's cottage, appropriately named Bleak House.[16]

One old place name which *doesn't* indicate the ghostly presence of a former wood is the misleading 'Forest of Dartmoor'. 'Forest' in this sense simply means a royal hunting ground: the Forest of Bowland and Forest of Exmoor are similarly treeless expanses. Dartmoor earned this title after it was acquired by the Crown in 1337 as a deer chase for the heir to the throne, the Prince of Wales.[17] Today, most of Dartmoor still belongs to Prince Charles's Duchy of Cornwall. The last deer on the moor was hunted to death around 1780, exterminated by the Duke of Bedford's hounds. Though few deer now exist to nibble fresh saplings, sheep have taken up the challenge instead.

But the word 'Dartmoor' itself contains an ancient, buried reference to woods. As the poet Alice Oswald reminds us in her epic poem of that name, 'Dart' comes from Brythonic Celtic for 'oak' (*derw*).[18] So the River Dart was, to the Celtic Britons, the river along which oak trees grew. Along much of it, they still do: from Wistman's Wood, lying on the upper reaches of the West Dart; down past the lichen-clad trees of Dartmeet, where the East and West Dart rivers join; via Holne Chase, a great secluded promontory of oakwoods, girdled by the river; past the Dartington Estate, pioneers of modern forestry; and out to Dartmouth and the sea. But pan back and out to the great denuded expanse of Dartmoor and you're reminded once again that, where oaks once grew, there's now mostly just wet desert. How has a landscape so rich in arboreal place names and folklore ended up so bereft of trees?

Realising that pre-Roman Dartmoor might have supported more temperate rainforest than today made me wonder something else. The Celts, who inhabited Britain 2,000 years ago, prior to the Roman conquest, had a caste of magician-priests called 'druids'. Legends about the druids associate them closely with sacred oaken groves, in which they gathered mistletoe and performed mysterious rituals. So, might old place names that refer to druids also point to the ghosts of long-lost woods?

Most of what we think we know about the druids nowadays is enjoyable rubbish: from the sickle-wielding druid Getafix in the *Asterix* comics, to the woad-soaked crazies in Jez Butterworth's fantasy drama series *Britannia*. A lot of what we're told about Celtic religion was just made up by eighteenth-century antiquarians, and later embroidered by modern pagans. But we do at least have some historical evidence that confirms the link between druids and the oak tree. As the Roman author Pliny the Elder, writing about the Celts in his *Natural History*, says: 'The Druids – for that is the name they give to their magicians – held nothing more sacred than the mistletoe and the tree that bears it, supposing always that tree to be the *robur* [*Quercus robur*, oak]. Of itself the *robur* is selected by them to form whole groves, and they perform none of their religious rites without employing branches of it.'[19] What's more, the word 'druid' is itself thought to be derived from the Celtic for 'oak' (*derw-*) and 'to see' (*-weid*); so 'druid' means 'oak-seer'.

After learning this, I got excited when I stumbled upon a hamlet on the edge of Dartmoor called, simply, Druid. Was Dartmoor home to the ancient Celts? But the hamlet, and the nearby Druid Plantation of beech trees, only date to the eighteenth century and were most likely so named by an excitable antiquarian who lived in the area.[20] Crossing's *Guide* states bluntly: 'Not only is there no proof that Druids ever were on Dartmoor, but some evidence that they were unknown in Devon.'[21]

At the risk of crossing Crossing, there *is* actually some evidence that druids and their sacred groves were known in Devon, pointing the way to another set of ghost woods. The Celtic word *nemeton* refers to a sacred grove or site. Just a few miles outside the northern edge of Dartmoor National Park, there is a cluster of place names containing 'nymet', 'nympton', 'nymph' and other corruptions of this same etymology. The villages of Nymet Rowland, King's Nympton, Bishops Nympton, Nymet Tracey; roads and paths still bearing names like East Nymph Lane and George Nympton Cross; and at least two Nymet Woods.[22] Most occur along the River Yeo, which used to be called the Nymet. Perhaps the river itself was the sacred site, but in pre-Roman times it was likely blanketed more continuously by oakwoods dripping with mosses and lichens. Some have suggested that 'Nymet' was the name of a forest through which this river once flowed.[23] As the historian Ronald Hutton says of the Celts, it's 'extremely likely that the British . . . continued to find a sense of the sacred in atmospheric parts of the natural landscape'.[24] What place could be more fitting for sacred rituals venerating nature than a temperate rainforest?

Spool forward 400 years from the Roman conquest of Celtic Britain to the fall of Rome and the arrival of the Anglo-Saxons, and we find other place names suggesting ghost woods in the landscape. In Anglo-Saxon, the word *bear*, and its various corruptions like *beare*, *bere*, *beer* and *burra*, all mean 'wood'. A surviving example is Black Tor Beare, another upland fragment of temperate rainforest that, mirroring Wistman's Wood, clings to the slopes of the West Okement in north Dartmoor. Burrator reservoir to the south, its watershed still peppered with lichen-clad oaks, is another remnant. At Beara Common, near South Brent, a scrap of common land is slowly being reclaimed by the scrub and trees that presumably once dominated it. But *bears* also crop up in places where the woods have long since disappeared: the farmsteads of Beardon to the west, and Frenchbeer to the east; the villages of Bere Alston and Bere Ferrers in the Tamar valley. All of these names could indicate the survival of temperate rainforest in these places up until the Norman Conquest. On Dartmoor, searching for lost woods means you're literally going on a bear hunt.

In search of ghostly *bears*, I had a look at some of the old tithe maps that exist of Dartmoor. These maps were drawn up in the 1830s when the Church still levied taxes on farmers, a hangover from feudal obligations in medieval times. Maps were created for each parish, down to an incredible level of detail: each field charted in order to work out how much money was owed to the Church. Crucially, the old field names were recorded – from an era when farmers would know their lands by a descriptive name, rather than simply a number. I searched the parish maps covering Dartmoor for field names that might contain clues for former woods. Sure enough, near the hamlet of Hexworthy, hugging close to the West Dart River, I found a cluster of three fields called The Bearas. Sloping steeply towards the rushing Dart, the cultivation of these fields even in the 1830s was recorded as being 'coarse'; today, Google Maps shows them to be scrubbing over with gorse and saplings. I visited them one bright spring day, and found them awash with blue-bells and bracken, indicator species of former woodland. Surely these *bearas* would once have been as wooded as the lower reaches of the Dart.[25]

Black Tor Beare itself, just fourteen acres in size today, was also once larger than it is now. According to Crossing, 'documentary evidence exists showing that this wood was once very much more extensive than at present: it probably stretched from the Island of Rocks [about 400 metres to the northwest] into the Forest'.[26] If so, Black Tor Beare in the past may have reached along the West Okement River for over two kilometres, twice its present length. Some indication of why the wood has since shrunk comes from old court cases dealing with the misde-meanours of commoners. In 1587, one William Bowden was prosecuted for cutting down certain oak trees at 'Blackerters Beare' in contravention of the Forest Law prohibiting the taking of 'green oak'. Another document from 1618 suggests that a black market in firewood had developed, referring to the payment of forty shillings for '8 acres of underwood growing within the forest of Dartmoor in a certain place called Black Tores Beare'.[27]

A final pointer to ghost woods on Dartmoor may lie in the 'Green Man' carvings found in many churches. These mysterious grinning heads, sprouting oak leaves and other foliage from their mouths, are a phenomenon dating back to medieval times, but their meaning remains hotly contested. In the 1930s, the folklorist Lady Raglan speculated that the Green Man was a symbol of an 'unofficial paganism' that had survived alongside Christianity, a peasant 'cult' that the governing clergy had failed to stamp out.[28] It has since become an icon of modern environmentalism, an image of humanity's dependence on nature. More recent scholarship has poured cold water on the idea of the Green Man as a pagan relic, but its presence may tell us something else instead.[29]

Some argue that the Green Man is a half-memory of the pre-Christian Wildwood; others contend that it appears in the greatest concentrations where there are surviving stretches of ancient woodland.[30] The church of St Pancras at Widecombe-in-the-Moor is one of nine on Dartmoor to boast Green Man carvings. Its fifteenth-century roof bosses contain no fewer than four Green Men, a depiction of the Arthurian hero Sir Gawain and his opponent the Green Knight, and numerous other carvings of trees and herbal cures. The chancel ceiling blossoms with leaves of oak, poplar and mulberry, and flowers of tormentil and hepatica.[31] If nothing else, such carvings point to how familiar medieval culture was with the local ecology: an agrarian society living in tune with the seasons, utterly dependent on nature for its food, fuel and medicines. This leads us to the most likely explanation of the Green Man's true meaning as a symbol of nature's regenerative power, and its likely co-option by the Church as a metaphor for Christ returning from the dead.

Dartmoor may be largely treeless today, but the evidence points to it once being home to extensive temperate rainforest. Hints lie scattered across the moor like clues at a crime scene: in old place names, in peat cores, in the distribution of woodland indicator species like bracken and bluebells. As the ecologist Dominick DellaSala observes, writing about the work needed to restore rainforest fragments in Europe: 'those undertaking rescue efforts today operate much like detectives in search of clues'.[32]

But if so much of Dartmoor's rainforest has disappeared, why has Wistman's Wood survived? Indeed, the grove harbours a still deeper mystery; one that Sherlock Holmes would have called 'a three-pipe problem'. A forensic scientist performing an autopsy on this fragment of forest would be forced to conclude that it has staged a miraculous resurrection. So why, since Victorian times, has Wistman's Wood roughly doubled in size?

Wistman's Wood has frequently been described as a moribund anachronism: a primeval throwback, left over from the last Ice Age; still surviving, but lacking vitality.

'One aged wood / Alone survives', mourned the Romantic-era poet Noel Carrington in his 1826 work *Dartmoor*, calling Wistman's a 'solitary wreck' that 'sleeps in the sunshine' and 'silently decays'.[33] Crossing says the wood is 'often spoken of as a relic of the ancient forest'.[34] Interwar botanists Miller Christy and R. H. Worth described the wood's trees repeatedly as 'dwarfed', 'deformed' and 'decaying'.[35] The first official guide issued by Dartmoor's then recently created National Park Authority in 1957 referred to the three upland oak copses as being 'stunted and fantastically gnarled', and representing 'parts of the indigenous woodland of the moor'.[36]

Many Edwardian visitors to Dartmoor, Conan Doyle included, were fascinated by its ancient Bronze Age settlements, ruins of past civilisations still present in the landscape. Hollywood's 1939 adaptation of *The Hound of the Baskervilles*, starring Basil Rathbone, describes Dartmoor in its opening title sequence as a 'vast expanse of primitive wasteland'. No doubt some saw the remaining scraps of upland oakwood in this same light: as part of a land that time forgot, a lost world still surviving whilst the rest of the world has moved on.

But, in truth, Wistman's Wood isn't a moribund relic at all – it's a living, growing, regenerating rainforest. 'It's like something out of *Ferngully*,' enthused one rambler I spoke to, referring to the popular 1990s animation about a rainforest. The wood is a National Nature Reserve, listed on the UK government's website. The entry casually records 'the woodland area doubling in size in the last 100 years'.[37] Yet no explanation is given for this astonishing, little-known fact.

To get an answer, I dug deeper. Rifling through books on Dartmoor, I found a few tantalising references to the wood's recent regeneration. The original Collins New Naturalist guide to the moor says that 'until the early 1960s Wistman's Wood tended to be regarded as moribund, at best unchanging, and possibly undergoing slow deterioration', before admitting it to be 'far from moribund now' and 'growing quite vigorously'.[38] The author of the most recent Collins guide to Dartmoor, Ian Mercer, goes further: 'There is photographic evidence . . . that Wistman's Wood had begun to extend itself in the twentieth century.'[39]

Photographic evidence! Reading this made me shiver. I felt like a consulting detective unearthing a cold case, and the coroner had just handed me a vital clue. So where were these photos?

My search took me first to the archives of the Dartmoor Trust. The earliest photograph they hold of Wistman's Wood is a faded black-and-white silver halide print, taken by the local antiquarian Robert Burnard.[40] It shows a solitary oak, stunted and gnarled, its compressed canopy and inky-black trunk resembling some sort of monstrous mushroom. Burnard's spidery handwritten caption reads: 'Typical oak tree of the lonely wood of Wistman, July 1888.'

Later that same year, Jack the Ripper would terrorise London. The conjunction of dates made me think of the sepia-tinted crime-scene photos taken by the hapless detectives of the Metropolitan Police as they hunted the Whitechapel killer. *Monochrome corpses. Moribund woods.* Across that gulf of time, no one has yet solved the Ripper murders, but a rainforest has doubled in size. And where countless Ripperologists have failed, the mystery of Wistman's Wood has been resolved by a group of ecologists.

In 1980, three plant scientists with excellent surnames – Molly Spooner, her husband Malcolm, and Michael Proctor – published a paper titled 'Changes in Wistman's Wood, Dartmoor: Photographic and other Evidence'.[41] The paper compared dozens of old photos of Wistman's Wood, such as those taken by Burnard in the late nineteenth century, with images from a hundred years later.

The differences are startling. A series of side-by-side pictures show trees, once dwarfed and shrunken in Victorian times, now tall and luxuriant with new growth. Old photos of empty boulder fields contrast with modern pictures showing them bursting with young saplings. A panorama of the whole wood from the early 1900s shows still more starkly what has happened since: Wistman's Wood has run wild, spreading down towards the Dart and clambering up the hillside towards the summit. 'It became evident that the wood has put on much new growth during this century,' observed the authors.

In a satisfying bit of sleuthing, the ecologists matched up old photos with present-day locations by using the unchanging boulders beneath the trees as reference points. From this they could gauge how far Wistman's Wood had grown, and even see changes in the size of individual trees. 'It is now clear that the impression of decay was deceptive,' they declared; 'rather than losing ground, the wood has expanded substantially.' Comparing Ordnance Survey maps of the wood from the 1880s to modern measurements by the Nature Conservancy Council, they found Wistman's Wood to have doubled in size from four to eight acres.[42]

The three ecologists were able to supplement their photographic evidence with measurements of tree heights going back centuries. The earliest historical mention of Wistman's Wood is by one Tristram Risdon in 1620. He wrote of the stunted trees that 'no taller than a man may touch to top with his hand' – that is to say, none of the oaks appeared higher than seven or eight feet. By the late Victorian period, various sources were stating the trees to be an average height of ten to twelve feet. By the time professional ecologists started measuring the trees more accurately between the two world wars, the tallest tree in the wood had reached twenty-six and a half feet. The most recent recorded height in the Proctor and Spooner list is a lofty thirty-four feet.

Why had Wistman's Wood put on such vigorous growth? One factor may have been a changing climate. Between the late medieval period and the mid-nineteenth century, Europe was in the grip of the 'Little Ice Age', a period of regional cooling during which the Thames repeatedly froze over and fairs were held on the ice. 'The conclusion is inescapable that in the early 1800s the condition of the trees was at a low ebb,' wrote Proctor et al. Yet, they added, 'climate cannot be the whole cause of the changes'.

Rather, there was also another culprit: a suspect with a decidedly sheepish appearance. 'Grazing is the major factor limiting tree growth over much of upland Britain, and has undoubtedly had important effects here', the ecologists concluded. Sheep, with their sharp front teeth, will readily devour saplings and prevent the regeneration of a wood. Cattle will also happily browse the low-hanging branches of trees. A reduction in the number of livestock grazing in and around Wistman's Wood during the nineteenth and early twentieth centuries could well help account for the wood's remarkable resurrection.

Sheep are also the prime suspects for why the wood has ceased regenerating in more recent decades. A study led by ecologist Edward Mountford in 2001, revisiting the earlier findings, found that 'grazing, browsing and destructive debarking . . . reduced rowan, eliminated holly, and prevented expansion after 1965'. Before that date, young saplings 'had almost broken free of check by browsing, but they have since been eaten back due to increased grazing'.[43] Despite having made a remarkable recovery, Wistman's Wood is now back in the intensive care ward.

Livestock, of course, belong to humans, who determine their numbers and where they graze. Rather than blame sheep per se, the true culprit is more likely to be the landowner. So, whodunnit?

It was Hallowe'en; the weather forecast was predicting torrential rain; and so, naturally, I was trying to persuade my partner Louisa to go wild camping north of Wistman's Wood. What better way to spend the spookiest night of the year, I reasoned, than in a tent near the most haunted wood in Britain, scaring each other witless with ghost stories whilst the wind blew a gale outside?

Well attuned to my whimsies, and despite her better judgement, Louisa agreed. As we set off for the moor, Dartmoor's notoriously changeable weather even lifted for a while, bathing us in sunshine when we passed along the footpath that leads through the boulder-strewn centre of Wistman's Wood. Every dripping branch was outlined with a golden green halo, as the sun caught the mosses and shone through the translucent polypody ferns. One old oak was so hung with string-of-sausages lichen that it resembled the beard of a venerable druid, its remaining autumnal leaves clinging to its crown like an oaken wreath.

Despite the time of year, the footpath through the wood was

surprisingly busy; clearly other people had had the same idea. We noted with annoyance the spiral moss-carvings that one visitor had made in some of the sphagnum-coated boulders. What had presumably been intended as some sort of offering to the pagan gods was, in reality, only damaging the plants. But though the trampling of visitors undoubtedly has some impact, the biggest threat to the future of Wistman's Wood is the grazing livestock. I searched the wood in vain for any signs of young saplings. Without this hope of renewal, the wood is dying on its feet.

Yes, Wistman's Wood has doubled in size over the past century. But it's still absolutely tiny: just eight acres, about the same area as the Oval cricket ground in London. And with renewed pressure from overgrazing, it too could one day succumb to the fate of Dartmoor's other ghost woods: remembered only in name. The history of the moor should serve as a cautionary tale to those assuming nature will inevitably spring back regardless of what we throw at it.

We ruminated over this fact as we hiked further to the north, to where the National Park Authority's online map had told us it was permitted to wild camp. Dartmoor is, in fact, the only place in England and Wales where it's legal to go wild camping: a quirk of history, resulting from the Dartmoor Commons Act of 1985 and the moor's centuries-old status as common land. Commons are where farmers are allowed to graze livestock, even though they don't own the land outright or hold leases on it. There are hundreds of Dartmoor commoners who hold common grazing rights to the moor. This ancient, complex system is fraught with politics: who gets to graze their livestock, at what stocking density, and with what level of impact on the moor, is a question that's being constantly fought over by commoners, conservationists and landowners.

Most of Dartmoor is owned by the Duchy of Cornwall, and Wistman's Wood lies in the centre of the Duchy's Dartmoor landholdings. But whilst much of Dartmoor is common land, the area surrounding Wistman's Wood isn't. Louisa and I discovered this ourselves when we crossed a wall about a kilometre to the north of the wood. Beyond it lay common land on which we could camp. Behind us, back towards Wistman's twisted oaks, was enclosed land.

The enclosure of Wistman's Wood – and another 20,000 acres of Dartmoor – resulted from the madcap scheme of Sir Thomas Tyrwhitt, a steward of the Duchy during the Napoleonic era. In my head, I imagine Sir Thomas to have been someone like Sir Edmund Blackadder: a long-suffering manservant to the foppishly indulgent Prince Regent, endlessly pursuing 'cunning plans'.

Tyrwhitt's cunning plan was to try to develop Dartmoor. To this end, he took advantage of the new turnpike road recently built across the moor – a kind of trans-Amazonian highway slicing through Dartmoor's wild heart – to encourage new settlers to the area. The 'improvement' of low-productivity moorland was all the rage in the late eighteenth and early nineteenth centuries, and heavily bound up with the movement to enclose commons. On Dartmoor, enclosures are called 'newtakes' – areas newly taken from the old commons, many of them at the behest of Sir Thomas. The grass is often greener on the inside of these fences, artificially so from the reseeded grasses and nitrogen fertilisers, but the improvement is measured in agricultural yields, rather than an improvement for nature. Wistman's Wood is inside the Longaford Newtake: the stone walls form a box dividing the wood from the West Dart River, from other enclosures to the east and south, and from the commons to the north.[44]

It's actually a wonder that Wistman's Wood wasn't 'improved', too, cut down to make way for something more efficient. In fact, the woods have had to endure all manner of industries over the centuries: coppicing; tin-miners hungry for firewood;

artificial rabbit warrens, built for fur-farming; and visitor pressure, from Victorian tourists to the lockdown crowds of today. But more than all of these, the waxing and waning of livestock grazing has had the greatest impact on the wood's fortunes.

Ultimately, Tyrwhitt's hopes of 'improving' Dartmoor were dashed upon its inhospitable climate and geography. One of the old stories about the moor carries a moral about the human folly of trying to dominate the natural world for private profit. It concerns Old Crockern, an ancient pagan deity said to inhabit Crockern Tor just to the south of Wistman's Wood: 'grey as granite, and his eyebrows hanging down over his glimmering eyes like sedge, and his eyes as deep as peat water pools'. The story goes that a rich farmer came to Dartmoor seeking to enclose and cultivate the moor, and was cursed by Old Crockern: 'If he scratches my back, I'll tear out his pocket.' Sure enough, the farmer's efforts to scratch a profit from the moor by draining the bogs only resulted in him draining his purse.[45]

The story remains a salutary tale for the Duchy's land agents and tenant farmers today. Nowadays, the Longaford Newtake is leased by the Duchy to Frenchbeer Farm Limited – a family business with various tenancies across eastern Dartmoor which has specialised in organic turkeys, but which still grazes livestock in Wistman's Wood and the surrounding enclosure. Decisions about how much grazing goes on in this newtake ought to be easier than the vexed politics of the surrounding commons, because the conversation is only between landlord and tenant, rather than dozens of squabbling commoners.

Could Wistman's Wood one day be allowed to expand into the surrounding newtake, by removing the sheep that currently nibble its young saplings? Surely the Duchy would allow it, with the green-minded Prince Charles at the helm, or his equally environmentally conscious heir, Prince William? Later, after our camping trip, I decided to ask them. I contacted the Duchy's head ecologist to enquire what plans it had for the future of Wistman's Wood. 'Conversations are happening,' I was told. The Duchy appeared open to the idea of allowing the wood to regenerate, and I came away from the call feeling excited that things were about to shift. But as time passed, my hopes began to fade. Though I had several pleasant follow-up calls and meetings with the Duchy to discuss the matter again, nothing seemed to be changing on the ground. We think of the trees in Wistman's Wood as being slow-growing, but perhaps our institutions are slower still.

The fate of Wistman's Wood lies in the hands of the Duchy and its tenant farmer. Will they continue to let the wood and its surrounding land be overgrazed? Or will they allow the ancient fragment of rainforest to regenerate, spilling over its present boundaries and recolonising the rest of the valley? These aren't just questions that apply to one small wood in Devon. They also apply to all of Britain's surviving temperate rainforests.

That Hallowe'en night, Louisa and I camped beneath a full moon. Wistman's Wood lay behind us, and the empty expanse of the moor stretched out ahead. We heated water on our tiny gas stove and gratefully gulped down the resulting lukewarm pot noodles, clasped in our shivering hands. Inside the tent, curled up in our sleeping bags, we told each other ghost stories: about wisht hounds, pixies, and the Wild Hunt. Outside, the wind began to howl. It was only later that we discovered the tent didn't do so well in gales. Poor Louisa awoke to find the tent fabric billowing back and forth inches from her face, battered by the storm. Neither of us slept well that night.

Bleary-eyed on our way back the following morning, we paused to shelter from the pouring rain in Wistman's Wood. We stared out to the west, across the Dart and the enclosure wall to the hillside opposite: a wet desert of purple moor-grass, studded with bleating sheep. On my Ordnance Survey map, I glanced at the name of the hill. *Beardown Tor. Bear* . . . the Anglo-Saxon for 'wood'. Once, in all probability, this whole valley would've been carpeted in rainforest. We'd come looking for ghostly hounds, but ended up finding a bear.[46]

3

Trespassing Botanists & Fern Maniacs

It was a rainy spring day when a group of us went trespassing in a rainforest.

Halfway up the River Dart in Devon, the river arcs north, bending sharply around a knuckle of land, before rushing south again. The raging waters make it a popular spot for daredevil kayakers. This promontory is called Holne Chase, and it supports a rainforest that's completely off-limits to the public. There are no rights of way over it, and the single footbridge across the river has a padlocked gate, its use reserved for the landowners. My discovery that the estate was registered offshore only added to the sense of mystery.

On previous visits to the area, I'd stared across the Dart to the inaccessible rainforests of Holne Chase, longing to find out what lay within them. It seemed to me like the Lost World of Sir Arthur Conan Doyle's novel of that name: a secret plateau, home to a wealth of forgotten species. So channelling my inner Professor Challenger, I decided to organise an expedition. I assembled a band of friends and intrepid adventurers: my old mate Nick, a seasoned trespasser; Robin, a biologist and climate campaigner; Lewis, an ecologist who's worked for the Field Studies Council; and Nigel, an expert in soil science. Together with myself and Louisa, this motley crew set off on our mission: a trespassory botanical survey of a secret rainforest.

Trespass: the word itself still carries a sense of jeopardy to me, a feeling of transgressing the rules. *Forgive us our trespasses*. But if you want to see much of the English countryside, you need to trespass. We still only have a Right to Roam over 8 per cent of England; over the other 92 per cent, the law of trespass still reigns supreme.[1]

A recce the weekend before had failed to find a promising route into the Chase. Our group had swapped increasingly wild ideas over WhatsApp. Lewis had even suggested bringing a rope in order to ford the river, but with the Dart in spate that week, following torrential rain, we all agreed that would be foolhardy. After a rendezvous at New Bridge car park, exchanging pleasantries between friends old and new, we began walking up the narrow country road that runs along the southern edge of Holne Chase, looking for ways in. It turned out to be surprisingly easy. Nigel's son Ed, back home from studying environmental law at university, had come along for the day, and had scouted ahead for entry points. A quick hop over a low stone wall, a shimmy under some barbed wire, and we were in. I was mildly disappointed: I'd been expecting something a bit more *Raiders of the Lost Ark*.

But if the trespass had made for an anticlimax, the rainforest itself did not. Sunlight glowed through the fresh spring leaves of oaks. 'They're almost acid green,' remarked Nick, looking piratical in his usual attire: a red neckerchief, a gold earring. We stopped at points to marvel at the richness of the flora growing on the forest floor. Nigel, silver-haired and avuncular, would help us identify the species: *cow-wheat, wood spurge, yellow archangel*.

As an amateur naturalist, I was glad to be guided by his expertise. Since moving to Devon, Louisa and I had been trying to learn the names of common plants, repeating them to each other as we spotted them in hedgerows and on road verges. As my urban eyes grew used to the dazzling variety of species abounding in the British countryside, trying to distinguish between the subtleties of leaf shape and petal colour became overwhelming, even offputting. I've never thought it necessary to be a botanical expert to appreciate nature. But the more I came to recognise the species I was seeing, the richer the experience became, like greeting old friends. During one of the lockdowns of the Covid-19 pandemic, a group calling themselves 'rebel botanists' had chalked pavements in cities with the names of the plants sprouting through cracks in pavements, helping to re-enchant a wider public with the botanical diversity around them.

'Amateur' is often used disparagingly, to dismiss ignorance, but the root of the word is Latin for 'love'. If we have no love for nature, it seems unlikely that we'll protect it. And without experiencing nature, regularly and close up, it's hard to see how that love can be cultivated.

'It's mad that all of this is shut off to the public,' Lewis said, shaking his head at the PRIVATE PROPERTY signs that had been nailed to trees along the route we took.

It hadn't always been this way. In Victorian times, top-hatted gents and ladies in crinolines would visit Holne Chase in horse-drawn carriages, taking the air along the woodland drives where we now trespassed.[2] Before that, it had been a deer chase, a place for the lords of the manor to hunt and stock up on venison. A solitary deer leapt across our path during our visit to this wet wilderness; in truth, I think I was more startled than it was. There's still something about trespassing that makes me nervous: a fear of confrontation. 'It's being discovered that makes me anxious,' agreed Louisa. 'Like Peter Rabbit being caught by Mr McGregor.' Though in the end we encountered no one, this rainforest was to us a landscape of fear, as well as of beauty. And whilst I resented Holne Chase now being a private rainforest, I was also aware that its seclusion might have contributed to its preservation.

Within half an hour of entering the forest, rain had started falling heavily, and soon we were all soaked to the skin. We followed a forestry track that spiralled up towards the apex of the Chase, and emerged into a glade that was awash with bluebells. They spread in drifts, breaking against the mossy trunks of oaks like waves upon the shore. Up ahead, the bluebell sea undulated downwards before rising again, following the contours of a ditch and earthen bank: the remains of a Bronze Age hillfort, dug some 5,000 years ago. We sheltered from the rain in the centre of the fort beneath the oak canopy, sitting in a circle, hungrily eating cake and cashew nuts and drinking cups of lukewarm flask coffee.

There was one more place I was keen for us to explore. As the Bronze Age hillfort showed us, Holne Chase had clearly been inhabited for millennia; but previous generations of humans had shared this place with species that were now long gone. Our Ordnance Survey maps pointed to one such species: Eagle Rock.

A scramble back down the steep hillside, cloaked in oaks, brought us to the churning waters of the Dart. Eagle Rock, an intimidating series of jutting crags, loomed above us like a fortress. There were no signs of any eagles here; but something else, almost as rare, caught my eye. The overhanging ledge of Eagle Rock was running with water; raindrops dripped continually from the spongy mosses and clumps of woodrush which clung gratefully to every available surface. And the wettest rock face – a damp cleft not far short of a waterfall – was carpeted in a vast expanse of tiny, exquisite filmy fern.

Filmy ferns are excellent indicator species for temperate rainforest. They thrive only in the moistest, most fecund conditions. They're much smaller than more common ferns like bracken: a frond of filmy fern is about the size of your thumb, with a membranous surface only one cell thick. In their natural, wet environment, they glisten like seaweed in a rockpool – but hold one up to the light and they glow with an emerald translucence. Even on damp Dartmoor, it's rare to find them.

Minutes later, I would discover something rarer still. Down by the river's edge, on a bend in the Dart where the waters roiled with foam-flecked waves, another much larger species of fern was growing in profusion. The fronds stood tall and ramrod straight, almost like bamboo, unfurling at their apex like a bishop's crozier. I called out in delight to the rest of the group. This was royal fern, *Osmunda regalis*, the crowning glory of Britain's rainforest ferns. Unlike the miniature jewels of filmy ferns, royal fern is one of our largest fern species, and can grow up to three metres tall.

Finding this spectacular fern growing here was particularly poignant, because royal fern was decimated during the Victorian period, leading to its decline across Britain. Many Victorian amateur naturalists didn't just enjoy looking for rare species; they felt the urge to collect them, too. 'Fernmania', as the nineteenth-century craze for fern-collecting was called, saw visitors to Britain's rainforest regions cut fronds for floral displays and even uproot entire ferns for transplanting into their gardens. It became not just a cultural fad but also an ecological assault on Britain's rainforest habitats.

Our trespass into this fragment of rainforest with its rare survivor species got me thinking. How do we get people interested in rainforest plants, many of which are seldom seen by the general public? And, having got the public interested, how do we prevent people from loving our rainforests to death?

This chapter is about the plants that characterise Britain's temperate rainforests: the ferns, mosses, liverworts and lichens that make them unique. It's about the botanists who study them, the herbalists who have made use of them and the artists who have admired them. But it's also about how our love of these plants has sometimes turned sour: about the people who have exploited them, and contributed to the destruction of their rainforest habitat.

First, a few key facts and terms that are useful for getting to grips with these organisms. Ferns, lichens, mosses and liverworts are sometimes referred to as *cryptogams* – plants and plant-like organisms which reproduce without flowers or seeds, using spores instead.[3] Turn over the frond of a polypody fern and you'll find it covered in orange dots called *sori*, each of which contains thousands of spores. The spore-producing fruit of lichens and mosses can be a bit harder to spot, but they're there nonetheless.

Mosses are more familiar to us, but less well known are their cousins, the liverworts. Mosses and liverworts differ in structure – liverworts tend to consist of lots of tiny lobes, whereas mosses consist of leafy stems – but they're sufficiently similar to be bracketed together as bryophytes. A biologist studies living things in general; a bryologist studies mosses and liverworts in particular. Birdwatchers call their hobby 'birding'; bryologists, adorably, call theirs 'mossing'.

Lichens are not, in fact, simply plants, but symbiotic organisms comprising both plants and fungi, bridging two entirely different kingdoms of life. Each lichen is an act of cooperation between an alga – a green plant generating food from sunlight using photosynthesis – and a fungus, which helps the alga anchor itself to a tree or rock and provides it with certain nutrients.

All of these organisms are incredibly ancient, and yet usually ignored. Ferns and mosses pre-date the dinosaurs, reaching back some 350 million years into the Devonian and Carboniferous periods.[4] Most of the world's coal originates from the primeval forests of these epochs. When Britain's coal pits were still operating, miners would regularly discover fossilised remains of ferns preserved in the coal seams. Lichens were once thought to be even older, representing the first emergence of plants on to land; and whilst they're now understood to be a mere 250 million years old, they were still colonising rocks and trees long before T-Rex walked the Earth.[5]

Despite the incredible antiquity of lichens and bryophytes, our culture still tends to overlook and denigrate them as being 'primitive', 'simple', or 'lower plants' – far less sophisticated than the modern flowering plants we fill our gardens with. As a result, comparatively few mosses and lichens have common names. Their identities are rendered remote by only possessing Latin names under the system of classification begun by the taxonomist Carl Linnaeus in the eighteenth century. Even Linnaeus had little time for lichens, dismissing them as the 'poor trash' of vegetation.[6]

Such prejudice is one facet of a wider disease of modernity: 'plant blindness'. Many people simply do not 'see' plants, other than as a blur of green. Some excuse this myopia as being biological rather than cultural, a function of how humans have evolved to keep an eye out for large animals that might be predators, or prey. Yet Indigenous cultures around the world invariably possess vast botanical knowledge because of the importance of plants and fungi for diets and medicine. Nowadays in the 'developed' world, however, we are far more cut off from nature. Environmentalists often have to drum up support for conservation initiatives by focusing on 'charismatic megafauna': lions and tigers and bears, oh my! Fluffy mammals and feathered birds are prioritised over scaly reptiles, slimy amphibians and buzzing insects. As for plants – they're usually at the bottom of the pile, or forgotten entirely.

But anyone who bothers to peer more closely at the panoply of lichens and mosses covering the trees in our temperate rainforests will discover entire hidden worlds. I've observed moss forests in miniature carpeting branches, and marvelled at veritable coral reefs of lichens suspended in the canopies of trees. As the moss expert Dr Robin Wall Kimmerer has written, 'the already gorgeous world becomes even more beautiful the closer you look'.[7]

What's more, Britain is home to hundreds of species of lichens and bryophytes. There are around 500 species of lichen in our Atlantic rainforests, according to the charity Plantlife.[8] The ecologist Derek Ratcliffe, who was the first to list all the species of Atlantic mosses and liverworts in Britain, counted 162 species. 'Although in many respects the British flora is poor by comparison with continental Europe,' wrote Ratcliffe, 'in its Atlantic bryophyte element, it is not only the richest part of the whole continent but it is also one of the richest areas in the world.'[9]

It feels to me that we need to become much more interested in the minutiae of the natural world around us. The little things are just as important as the charismatic megafauna. In fact, I would argue that Britain's rainforests contain numerous examples of charismatic *microflora*. One in particular has come to captivate my attention and excite my curiosity, to the extent that tracking it down has become an obsession. Dear reader, let me introduce you to the improbably named, but wonderfully charismatic, *tree lungwort*.

Tree lungwort is a glorious lichen of Britain's western rainforests. Its Latin name, *Lobaria pulmonaria*, sounds like a wizarding spell out of Harry Potter – which also perfectly describes its magical allure. It has certainly enchanted me. I first saw it growing on an old oak tree near Dartmoor. To my eyes, it looked like dragon scales; its grey-green foliage arching out of the trunk like the wings of the Jabberwock in Lewis Carroll's iconic poem. Its name derives, however, from a more gruesome resemblance: its blistered surface and apparent air-chambers bring to mind the inside of a lung.

Many plant species in Britain have common names ending

in '-wort'. It's a suffix that means 'worth', and comes from the plant having a perceived medicinal value. In the case of tree lungwort, it was once thought to help cure pulmonary diseases (hence *pulmonaria*), being prescribed to treat coughs, asthma and even tuberculosis. Medieval and early modern herbalists were besotted by the 'doctrine of signatures' – the idea that medicinal properties of plants were suggested by their shapes.[10] It's true that the way *Lobaria pulmonaria* branches out resembles the bronchial structure of lungs, and its mottled surface looks a little like alveoli. Its use for lung ailments goes back at least as far as the 1500s: according to one account, 'it was usually boiled with water or milk and drunk, or made into an ointment for external use'.[11]

An asthma sufferer myself, I've occasionally been tempted to give this witches' potion a try, but have thought better of it – because tree lungwort is today desperately rare. The Industrial Revolution was a disaster for many lichens, particularly *Lobaria*, which found itself choked by the rise in sulphur pollution and acid rain. Its survival in many of our temperate rainforests is therefore an indicator not only of a moist, oceanic climate, but also of very pure air. Sometimes referred to as the 'lungs of the forest', *Lobaria* is as sensitive to air pollution as we humans are. If we cleaned up our air, we'd likely see more lichens and fewer cases of asthma. (Harvesting tree lungwort for selling is also nowadays illegal under the 1981 Wildlife and Countryside Act.)

So although I had no intention of foraging any I found, I was still seized by a desire to find this magical species in the wild. My quest to track it down started out simply enough: I began by looking up. *Lobaria* flourishes on old trees where there's plenty of light, so on my walks in the woods I started staring up more at branches, squinting to make out the lichens. But whilst this no doubt earned odd looks from passers-by, I was at first disappointed. Lungwort, being uncommon, was going to be harder to find than I thought.

Over time my quest became more quixotic. I sought help, of a kind, from botanical experts. The National Biodiversity Network Atlas is a wonderful online resource containing hundreds of thousands of recorded sightings of plants, animals and fungi.[12] As a map geek, the Atlas had me salivating: here was my treasure map, with X marking the spot for every specimen of lungwort in Britain. All that I needed to do was load the app on my phone, put on my hiking boots and go find some. Or so I imagined.

One such weekend treasure hunt turned into an increasingly frantic search lasting for hours. A recorded sighting of lungwort on the NBN Atlas had caused me to drive fifteen miles to a wood on Dartmoor's edge, dragging Louisa along for the ride; and now the lungwort was nowhere to be found. We paced up and down the narrow forest road where the map claimed it to be, and circled the trunks of trees, gazing upwards. I began to think I'd embarked on a wild goose chase. Many of the entries in the Atlas are decades old, from a time before smartphone GPS allowed botanists to record sightings with pinpoint precision. Clearly, the map is not the territory. So, with even Louisa's patience wearing thin, I admitted defeat, and we began to walk back to the car.

On the brow of a hill, I paused to rest by a rowan tree, and stood awhile in thought.

> *And, as in uffish thought he stood,*
> *The Jabberwock, with eyes of flame,*
> *Came whiffling through the tulgey wood . . .*

'There it is!' I burbled, pointing at an old sycamore in the field ahead of us. Scaly lungwort erupted from its trunk. I leapt over the hedge to get a closer look. *O frabjous day!*

By now, I'd caught the bug. Seeing just one specimen was never going to be enough. Later that summer, visiting my parents in Cornwall's rainforest zone, I discovered some lungwort on a walk. But it was too high up in the branches of a veteran oak tree to get a good look at it. So I returned to Mum and Dad's, and after failing to persuade them to let me borrow a stepladder, hit upon the cunning idea of gaffer-taping my iPhone to a telescopic walking pole to make an arm's-length video camera. I shot some great close-up footage of the lungwort; but I also realised that I had now become lichen-mad.

In my defence, I'm not the only person to become captivated by the weird beauty of this lichen. Another fan was the German biologist Ernst Haeckel, who first coined the term 'ecology', and who, in the late nineteenth and early twentieth centuries, also made extraordinary prints of 'artforms in nature' – illustrating the amazing symmetry and geometric shapes displayed by jellyfish and fungi, shells and corals. One of Haeckel's prints has at its centre an illustration of *Lobaria pulmonaria*, its scaly wings starkly silhouetted against a black backdrop. It could be a botanical drawing of an alien lifeform, captured by an astrobiologist on a future survey of Mars. Yet we have this wonderfully exotic species in our backyard, here in Britain.

Lungwort is often confined to rare patches on veteran trees, but sometimes you can find rainforests full of the stuff. One such place is the Dizzard, on the north coast of Cornwall, blessed with a wonderfully Cornish name. My parents tipped me off about it, and we visited together, a family day out in a rainforest. The approach wasn't immediately promising: all that we could make out was the cliff edge and the sea, with nothing in between. As we got closer, we saw that the land didn't end at the clifftop, but in fact sloped sharply down towards the waves. Here, battered by the wind and drenched by sea-spray, a rainforest thrived. *An ocean of freshing verdure toss'd by gales Atlantic.* There was another surprise to come: the trees in this rainforest were barely taller than me.

The entrance to the wood felt like the mouth of a cave. We stood on its cusp, sunshine on our backs, the cool, dark interior of the forest tumbling away before us. I could sense my parents' slight trepidation. Louisa and I ducked beneath the overhanging branches and beckoned to them to follow. Dad, six feet tall, stooped under the low canopy; Mum trod gingerly over the slippery woodrush and brambles carpeting the forest floor, testing the path with her walking poles. All around us were tiny shrunken oaks: knock-kneed, their branches crooked like elbows, their trunks coiling like question marks. And on those trunks, *Lobaria* bloomed in abundance.

The Dizzard is thought to be thousands of years old. Its oaks appear to have been bonsaied through a combination of prevailing winds and thin soils: the escarpment on which it grows is a kind of slow-motion landslip. In places, the trees were so small that we could raise our heads above the forest canopy, feeling like giants in a Lilliputian forest. And though the understorey felt damp and sheltered, dappled sunlight still permeated through the low branches. This combination of conditions – light, moisture, clean air from the sea breeze – had conspired to create a paradise for *Lobaria*.

There was so much of it that we had to take great care not to knock it from the trunks. In summer, tree lungwort – like many lichens – dries out, becoming crisp and brittle to the touch. Alongside the oaks and hazels, Mum spied the white bark and maple-like leaves of a rare wild service tree, an ancient woodland indicator species – and therefore a good sign that this rainforest has been here for a very long time indeed.[13]

That's the other ingredient required for lungwort to flourish: time. It spreads very, very slowly. Lungwort's fruiting bodies, called *apothecia*, can distribute spores over some distance, but they only appear after about twenty-five years. It's also capable of reproducing asexually, like many lichens, via granule-like structures called *soredia*: essentially, a fragment of the lichen in miniature, capable of growing into a new organism. But soredia are only capable of travelling tens of metres, often relying on raindrops for dispersal, dripping from branch to branch.[14] Combined with *Lobaria*'s pickiness for where it grows – like with Goldilocks, everything has to be *just right* – and you can see why it takes a long time to get around. The legendary botanist Francis Rose categorised *Lobaria* as part of a 'climax community' of lichens, tending to only appear in old-growth woodlands that have gone through earlier stages of plant succession and reached a 'steady state'.[15]

Lungwort's rarity and slow reproductive cycle makes it something of a nightmare for conservationists. In 2020, a veteran oak tree in the Lake District sporting 'one of the single largest communities of *Lobaria pulmonaria* anywhere in England' blew down in a storm.[16] The oak tree's owners, the National Trust, launched an immediate rescue mission. They couldn't save the tree, but they could try to salvage the lichen – by translocating it to other trees. Lichenologists removed three square metres of lungwort from the fallen trunk, and reattached it to dozens of other trees in the same valley using staples, wire mesh and eco-friendly glue. It sounds slightly desperate, but it can work: over time, the lungwort can become established and reproduce naturally. (A conservationist once told me about a still more outlandish method for propagating *Lobaria*: by putting fallen pieces in a food blender, and spraying the resulting lichen soup on to branches with a super-soaker. Let's just say that the results are not yet in.)

If it's so hard to conserve, why bother? Because tree lungwort isn't just amazing to look at. In common with other species of lichens, it also forms an essential part of the wider rainforest ecosystem. Lichens like lungwort provide food and shelter for other species: they're eaten by everything from deer down to snails, and plucked from trees by birds to use as nesting materials. And *Lobaria pulmonaria* can also play an important role in a rainforest's nutrient cycle. When lungwort forms a three-way symbiotic relationship with a blue-green algae called *Nostoc*, it can fix nitrogen from the air, like a legume does. When the lungwort later falls to the forest floor, distributing its nutritious nitrogen, this evolutionary threesome adds to a rainforest's fertility.[17]

Though I continued to get a buzz from tracking down *Lobaria*, I slowly came to realise that focusing monomaniacally on one species has its limitations. First, tree lungwort is only one indicator of temperate rainforest, and not a wholly reliable one at that. It can also be found in some non-rainforest woods like the New Forest, though it has mostly retreated to the damp, clean air of western Britain.[18] And second, I learned that a better way of seeing the world through an ecological lens is to look not just for individual species, but whole *communities*.

A biological community is a set of interacting species that inhabit a particular location.[19] There are two main lichen communities that are particularly characteristic of Britain's temperate rainforests: the *Lobarion* community, and the *Parmelion*.

The Lobarion community takes its name not just from *Lobaria pulmonaria*, but from other species that often share the same ecological niche. *Lobaria scrobiculata* – or 'Lob scrob' for short – is lungwort's gunmetal blue cousin. Whilst it lacks lungwort's dragon-wing shape, it still arcs flamboyantly from the trunks of trees. *Lobaria virens*, meanwhile, is a startlingly green lichen with a shiny satin skin, which loves the dampest parts of the Lake District, Wales, Scotland and the Westcountry. Rarest of all is *Lobaria amplissima*, which hugs closer to tree bark than Lob scrob or lungwort, but wrinkles like parchment when dry. The more species of *Lobaria* you find, the older and richer the rainforest: in parts of western Scotland, I've been lucky enough to see all four species on one tree.

The Lobarion community also encompasses other species, too. The various species of *Sticta* lichens are particularly notorious:

not for their appearance, but for their smell. *Sticta sylvatica*, *Sticta fuliginosa* and *Sticta limbata* all stink like rotting fish when they're wet. What possible evolutionary purpose this could have is a mystery. A botanist friend told me that he once left some *Sticta* in his workplace overnight, and came back in the morning to find his colleagues searching high and low for the source of the stench.

More congenial to the nose and eye is felt lichen, *Degelia atlantica*, which I've only ever seen growing in dampest Scotland. It resembles a cluster of scallop shells, forming a velvety grey mat over branches. A more common set of Lobarion lichens are the *Peltigera* species, or dog lichens – so-called because of the white 'teeth' on the lichen's underside, resembling a dog's canines. Presumably because of this, *Peltigera canina* was once listed in old pharmacopoeias as a cure for rabies.[20]

Whilst the Lobarion community tends to favour trees with alkaline or neutral bark – old oak, ash, hazel, willow – the Parmelion community of lichens is able to thrive also on trees with more acidic bark, like birch, hawthorn and alder. In contrast to the larger-than-life personalities of the Lobarion community, members of the Parmelion can seem humble and shy, their smaller forms clinging closer to branches. But they, too, include many delights.

There's *Parmelia saxatalis*, a blue-grey lichen with the evocative common name of 'crottle': it was once used in Scotland to make orange dye for colouring Harris tweed.[21] Another Parmelion species formerly used for dyeing is *Ochrolechia tartarea*, a 'cudbear' lichen, whose Latin name hints at its appearance: it has ochre-coloured fruits that look like jam tarts. Or, as Louisa described it, 'like tippex has exploded in the bottom of your bag, and someone's scattered Haribo fried egg sweets on it.' *Sphaerophorus globosus*, meanwhile, is more commonly known as a coral lichen. I love the idea of Britain's rainforests supporting coral reefs, suspended high in the air.

Sometimes bracketed alongside these species are two of my favourites, which thrive particularly in the rainforests of the Westcountry. The 'string-of-sausages' lichen, *Usnea articulata*, hangs like an eggshell-green necklace off hawthorns and willows on Dartmoor. In some places, I've seen it growing so exuberantly that it's hard to make out the tree under the curtains of lichen. Up close, you can see that its branches form miniature sausages, strung together in a chain. My other favourite lichen of Devon is *Usnea florida*, or 'witches' whiskers'. Its disc-shaped fruit stare out like eyes on stalks, surrounded by whiskery eyelashes. You may not be looking at it, but it's looking back at you, giving you the evil eye. Collectively, the various species of *Usnea* are often called 'beard lichens', their straggly tassels transforming trees into bristly old wizards and witches.[22]

Given this association with sorcery, it's no wonder that lichens crop up in traditional herbal medicines, both in Britain and across the world. But it may be that there's more to lichens than just quack cures. As Dave Lamacraft, lichens expert at the charity Plantlife, notes: 'The old-man's beard lichen contains usnic acid, which is considered more effective than penicillin against some bacteria. We've only just touched on the capacity of these organisms.'[23] One recent review of the pharmaceutical properties of lichens concluded they represent 'an untapped source of biological activities of industrial importance . . . their potential is yet to be fully explored and utilised'.[24] Much has been written about the importance of 'rainforest medicines', and how the destruction of the world's tropical rainforests may be depriving humanity of natural cures for many diseases. Perhaps Britain's own rainforests also contain medicinal secrets yet to be discovered.

But it should be possible for us to appreciate rainforest lichens for simply existing, not just for their utility to us. Someone who did was the Victorian artist and early environmentalist John Ruskin. Ruskin's attention to detail was legendary. One of his most famous drawings is *Study of Gneiss* (1853–4), a hyperrealistic rendition of a single boulder, sketched in the ancient woodland of Glen Finglas in the Trossachs. Though clearly drawn to the spot by its geology, Ruskin gave just as much attention to the foliage, depicting in infinitesimal detail the numerous lichens, ferns and epiphytes clinging to the rock's surface.[25] Another of his meticulously observed works, *Study of Moss, Fern and Wood-Sorrel, upon a Rocky River Bank* (1875–9), almost certainly shows the tiny rainforest species filmy fern.[26]

It's perhaps no coincidence that Ruskin, as a fierce critic of Victorian industrialisation, was drawn to lichens, which suffered greatly from the sulphurous smogs and acid rain generated by Britain's burning of coal. With the country's ancient, tenacious communities of lichens forced to retreat to the remote west, where the air remained pure, Ruskin the eco-warrior leapt to their defence, calling them 'the most honoured of the earth-children'.[27] In a time of climate crisis and ecological collapse, lichens deserve to be honoured still more.

Artists, herbalists and ecologists have different ways of appreciating the living world. But to really get under the skin of our rainforest flora, I needed to meet some bryologists.

Alison and Ben Averis have literally written the book on Britain's upland vegetation. They co-authored *An Illustrated Guide to British Upland Vegetation*, the first comprehensive tome on plant communities in the British uplands, now used by ecologists as a standard reference work.[28] They're a botanical power couple, together having surveyed, classified and studied the bryophytes of Britain's temperate rainforests and upland heaths for over thirty years. To cap it all, Ben's an amazing illustrator, whose beautiful drawings perfectly capture the moody hues and ethereal beauty of our rainforests. I couldn't wait to meet them.

When Louisa and I eventually did so in the winter of 2021, it was in the most perfect spot imaginable: Glen Coe's Lost

Valley, deep in Scotland's rainforest zone. Ben and Alison live near Edinburgh, but had – with characteristic generosity – taken the time to drive the breadth of the country to spend two days with us. We greeted one another in a small car park by the side of the road. I recognised Ben from a recent Zoom call we'd had: bespectacled, his silvery-blond hair worn in a ponytail. Alison was slight, wearing a peaked cap to keep off the rain, a silver earring inscribed with Celtic patterning in one ear. I liked them both immediately.

All around us, Glen Coe sprawled in its magnificence. The series of vertiginous ridges known as the Three Sisters loomed overhead; the valley floor, carved by a glacier 10,000 years ago, conspired to make its human visitors feel insignificant. As we drank in our surroundings, Ben pointed out where we'd be heading: the wooded ravine of the Coire Gabhail, the 'Lost Valley', which cuts between two of the mountainous Sisters. 'We've surveyed all the vegetation in this valley, as far as the eye can see,' he said, softly and without boast. 'Today's a nice day,' observed Alison, 'but surveying on some of those slopes in horizontal rain for weeks on end could get fairly miserable.' I marvelled at their dedication, and how intimately they must know this landscape, better than any landowner.

Waterproofs on and boots laced, we set off up the corrie. As we climbed, the heather and bilberry bushes shaded into birch and rowan saplings, and the moorland grasses were replaced by mosses and ferns. Up ahead, Alison spotted something growing by the waterfall that cascaded through the ravine. Whatever it was clearly excited her interest, because she and Ben practically skipped over the rocks to get to it. A mass of orange-tinted liverwort sprouted from the base of a rowan, its trunk already covered in filmy ferns and bilberry plants growing epiphytically. 'This is an uncommon oceanic liverwort called *Herbertus hutchinsiae*,' explained Ben, when I asked him what all the fuss was about. 'Why do you like it so much?' Louisa wondered.

'Oh, because it brings back so many happy memories of days amongst damp vegetation!' beamed Alison.

We perched on boulders to eat our sandwiches. A jet-black raven started croaking in the woods above us, altering its pitch on every third croak. *Awk, awk, ARRK.* 'Quite a melodic range,' said Ben, drily.

We talked about how he and Alison had met. Ben had started out in art college, then went on to do bird surveys for the RSPB, before moving to study plants for a government body formerly called the Nature Conservancy Council (now NatureScot in Scotland). There he met Alison, who'd not taken Ben's roundabout route, instead knowing from a young age that she wanted to study the upland plants of Britain. A lot of their work is done together, symbiotically, but they've also carved out their own ecological niches: Alison's focus is on montane habitats, whilst Ben's love is for temperate rainforests.

Most people come to Glen Coe for a sense of the sublime: the grand vistas, the lofty peaks. It was the Romantics who first turned the British public on to the idea that nature at scale is awe-inducing. But to bryologists like Ben and Alison, it's the tiny things that are truly awesome.

Learning to see mosses and liverworts requires unlearning some of the assumptions our brains have made about the world around us. Our plant blindness reduces the dizzying diversity of bryophytes and dumps them all into one bucket labelled 'moss'. 'But we all manage to distinguish between tiny emoticons on our phones every day,' said Ben, reasonably. 'Why not with plants?'

Even so, I'll freely admit that it was hard going at first. Alison and Ben kept up a steady barrage of taxonomic names, unpretentiously but with a frequency that felt like rain. *Dicranum majus . . . Bazzania trilobata . . . Frullania tamarisci.* But as my eyes adjusted to this new reality, the monolith of 'moss' began to dissolve into a welter of new forms. *Breutelia* is a moss that resembles a bottle-brush, its bristles all pointing out from the central stem. *Ptilium*, an uncommon moss of northern woods, glows gold and is shaped like an ostrich feather.

At this level of detail, even bryologists sometimes need a helping hand. Ben lent us his hand lens, its 10× magnification making it easier for us to tell apart two very similar oceanic liverworts, *Plagiochila spinulosa* (spiny) versus *Plagiochila punctata* (spiky, like a punk's hair).[29] In a rainforest like the one in Glen Coe, the sheer dampness can also make it harder to see. As Alison and Ben stood with their faces pressed up against tree trunks, they would sometimes have to squeeze the rainwater out of the moss they were peering at to get a better look at its form.

Experiencing these luxuriant rainforest mosses and liverworts doesn't just rely upon sight. Robin Wall Kimmerer writes about how moss feels to touch, inviting us to 'run your fingertips over a silky drape of *Plagiothecium* and finger the glossy *Brotherella* brocade'.[30] I love the smell of moss, and find a visit to a temperate rainforest isn't complete without getting up close to a patch to breathe in its earthy, humus scent. But trained bryologists have developed a more discerning nose than mine. *Scapania ornithopodioides*, a liverwort found in submontane rainforests from Scotland to the Himalayas, smells to Ben Averis like unwrapping a present. It's a perfect metaphor for how he and Alison have come to view the world.

Temperate rainforests are full of gifts for those who visit them. The problems start when visitors arrive wanting to take home souvenirs. One particular episode in Britain's relationship with its rainforests offers some salutary lessons: the extraordinary, but almost forgotten, history of the great Victorian fern craze.

Between the accession of Queen Victoria in 1837 and the outbreak of the First World War, the British public were seized by a passion for ferns. It started out modestly, with plant collectors bringing back exotic tree-ferns from the far-flung corners of the British Empire to display in the Royal Botanic Gardens at Kew. But with the rise of the Victorian middle classes, and their interest in gardening and nature, came a growing fascination with ferns that fast became a cult. Something about ferns, their fractal patterns and lustrous greens, possessed the Victorian imagination. As the historian Sarah Whittingham records, 'the fern craze affected all classes of society, and had followers among men, women and children throughout the British Isles'.[31] In 1855, the Reverend Charles Kingsley, author of *The Water Babies*, coined the term for which this phenomenon became known: 'Pteridomania', or fern-madness.

What makes fernmania interesting, setting it apart from other nineteenth-century fads for plants, is that it was essentially an obsession with Britain's own rainforest flora. Colonialism and the international trade in exotic plants certainly fuelled it. But it was grounded in the ferns that grow naturally in Britain's damp climate: the sixty species native to this country, of which around fourteen are common in our temperate rainforest zone.[32] The Victorians fell in love with our rainforests, even though they didn't call them by that name.

Frond-fanciers were therefore increasingly drawn to the western fringes of Britain, where ferns grow most luxuriantly. Devon, in particular, became known as *the* Fern County. Charles Kingsley's sister Charlotte helped establish the lush Devonian valleys as the go-to place for fern hunters with her 1856 guide *Ferny Combes*, in which she described the county as 'rich in botanical treasures'. She visited Holne Chase – where we trespassed at the start of this chapter – and found the river, as we did, 'fringed with *Osmunda regalis* [royal fern] of great size and in marvellous profusion'.[33] The writer Francis George Heath called Devon 'Fernland', and rhapsodised about how its mild and wet climate led to an abundance of ferns: 'They clothe its hillsides and its hill-tops; they grow in the moist depths of its valleys; they fringe the banks of its streams'.[34]

Soon the new railway network was carrying frond-fanatics to every corner of western Britain in search of ferns. Much of the newfound interest reflected a growing appreciation of the natural world: this was the golden age of the amateur naturalist, when eccentric parsons and bored wives would spend their spare time writing treatises on rare butterflies. Guidebooks for fern identification became bestsellers. Thomas Moore's *A Handbook of British Ferns* (1848) went through numerous editions, and a folio edition pioneered a new form of 'nature printing' in which real fronds were pressed on to a soft lead plate to create photorealistic impressions.[35] Fern hunters searched not just for uncommon species of fern, but also for the many varieties – or natural mutations – that occur within fern species. Most sought-after was a variety of lady fern in which the fronds crisscrossed like a crucifix. For devout Victorians, the mark of the Creator could be found in the natural world.

Some sought to recreate the ferny habitats which they admired in their own homes and gardens. The 'Wardian case', a kind of glass terrarium, afforded owners the chance to grow ferns even in the smoggy cities of industrial Britain. Aristocrats with Westcountry estates could go further still, constructing not only glasshouses but outdoor ferneries, stuffed with fern species both native and exotic. The Lost Gardens of Heligan in Cornwall – a Victorian estate that fell into ruin before being restored in the 1990s – includes some magnificent fern-gardens, nicknamed 'The Jungle' and 'The Ravine', both of which make use of valley microclimates to encourage the growth of ferns. At Canonteign, in Devon, the landowner Lady Exmouth oversaw the creation of an artificial waterfall and extensive fernery, populated with tree-ferns from New Zealand as well as species growing naturally in the estate's damp gorges.

Fernmania spread into every facet of Victorian life, inspiring art, design and endless pieces of chintzy merchandise seeking to cash in on the trend. Fern designs appeared on garden furniture, fireplaces and crockery; sculpted fronds adorn the stone pillars in Oxford's Natural History Museum. Next time your grandma offers you a custard cream biscuit, take a closer look at it: its pattern of uncurling ferns is another legacy of the fern craze.[36]

Much about fernmania clearly sprang from a deep-seated love of the natural world. When I first read about it, I initially laughed, dismissing it as a typically Victorian fad. Then I thought about my own mad quest to track down rare species of lichen. I can certainly see where such obsessions arise from. But what I can't understand, still less condone, is how the fern craze went sour, leading to the destruction of the very plants that fern-fanciers claimed to love.

Because, for many fernmaniacs, simply observing and appreciating ferns wasn't enough: they had to collect them, too. Fern-collectors did immense damage to the rainforest habitats in which ferns thrived, not only cutting fronds for floral displays but also digging up entire plants for their gardens. They trawled the countryside for rare ferns with trowels and blades: gouging, digging, destroying. One fern-collector armed themselves with a fifteen-foot-long piece of bamboo with a knife tied to the end; another with a six-inch-long steel pick on the end of a walking stick.[37] I was reminded of when I gaffer-taped my phone to a pole; but at least I was only using it to take photos, not plants. After all, the ferns would often not even survive transplanting.

The Cornish botanist Frederick Hamilton Davey wrote despairingly: 'Such shameful plundering has gone on that I now hesitate to speak or write about localities where the royal fern grows.'[38] Another account from Devon bemoaned how '*Osmunda regalis* has been utterly eradicated from near Cornwood where it once grew with fronds six feet high', and lambasted the 'itinerant fern collectors who were exposing for sale, large mats of *Hymenophyllum* [filmy fern], torn ruthlessly from their rocks, the scars of which remain in the Meavy and Cornwood valleys to the present day'.[39]

Such devastation also spurred on a dawning environmental consciousness amongst some Victorians. The writer Nona Bellairs was ahead of the curve when she called for laws to protect ferns from collectors in her 1865 guidebook *Hardy Ferns*. 'The poor Ferns, like the wolves in olden time, have a price set upon their heads, and they in like manner will soon altogether disappear,' she wrote. 'We must have "Fern laws", and preserve them like game.'[40]

In the absence of such environmental legislation, the authorities sought to prosecute fernmaniacs using existing laws on theft and criminal damage. One sign put up by a local police force offered a reward of £1 – a princely sum in those days – for information that would lead to the conviction of 'any person or persons taking ferns'. In 1896, two fern-stealers were charged at Totnes magistrates' court with damaging Devon hedges, and sentenced to four and six weeks' hard labour each.[41] Some landowners sought to discourage bad behaviour by threatening to close off paths. A tourist guide to north Wales warned that the walks around a particularly ferny lake were 'open to the public on the express understanding that . . . [visitors] will not disturb or remove the roots of ferns'.[42]

Fernmania eventually led, remarkably, to one of the earliest environmental regulations in Britain. In 1906, Devon County Council passed a bye-law mandating that 'no person shall uproot or destroy any ferns or other wild plants growing in any road, lane, roadside waste, wayside bank or hedge, common, or other public place'. The penalty for breaking Devon's Fern Law was £5 – or over *£600* in today's money.[43]

By then, however, fernmania was all but spent, a fad that faded away with the passing of the Edwardian belle époque. The damage had been done.

Fernmania has now long since disappeared from British consciousness. Yet whilst destructive fern-collecting has thankfully been consigned to the dustbin of history, there are still plenty of signs of widespread disregard for the plant life of our temperate rainforests. I think back to the moss-carvings I've seen scored into rocks in Wistman's Wood and shudder. This is the dark side to amateur enthusiasm for nature: a love that is blind to the destruction it can cause, whether to prettify a Victorian garden or for a viral post on Instagram.

But a far more insidious threat to our rainforests, I'd argue – along with other precious habitats – is our culture's plant blindness. Too many of us are simply unaware of the wonders of the plant kingdom: the worlds in microcosm which thrive in our fragments of rainforest. I was, too, before I was lucky enough to move to a part of the country that opened my eyes to what I'd been missing out on.

Simply walling off our remaining fragments of rainforest habitat and hoping no one notices them doesn't seem to me to be a good strategy for protecting them. We need to combine public access with public education. The government ought to make the teaching of natural history compulsory at school, and give every schoolchild the chance to experience nature close up. When you visit one of Britain's magnificent rainforests, remember the essence of the Countryside Code: take only photographs, leave only footprints.

We can all learn more about the plants around us, whether from the chalk notes on pavements left by rebel botanists, or by reading one of the guides to our rainforest lichens produced by charities like Plantlife. For what we fail to perceive, we often fail to protect. As the scientist Robert Michael Pyle says: 'People who care conserve; people who don't know don't care. What is the extinction of the condor to a child who has never known the wren?'[44]

4

Seeing the Wood for the Trees

On a windswept hillside, high in the Lake District, is a rainforest not shown on any map.

Young Wood is the highest Atlantic oakwood in England, and – despite its name – is thought to date back to the last Ice Age, an untouched survivor of the once-mighty Wildwood. But there's no sign of it on the maps produced by the Ordnance Survey, and it's missing from the official record of ancient woodland. It's conspicuous only by its absence in the maps of woods kept by the Forestry Commission, or in the habitat maps published on the UK government's website. Even Alfred Wainwright, the inveterate rambler of Lakeland fells, failed to include it in his meticulously illustrated walking guides.

Yet it undoubtedly exists. And on a cold and misty day one January, I ventured to see this elusive ghost with my own eyes. Two friends at the RSPB, Lee Schofield and David Morris, offered to guide me to the wood. Lee, tall and lean with salt-and-pepper hair, has managed the RSPB's Haweswater nature reserve for over a decade; whilst David, a blunt-speaking Yorkshireman, is responsible for all of the charity's sites across Cumbria and northeast England. Lee calls Young Wood 'one of the Lake District's most fascinating ecological fragments'. We set off from the village of Mungrisdale on the eastern edge of the National Park, following the footpath west into the cloud-capped dome of the Skiddaw massif.

Ahead of us lay a glacial valley leading to Bowscale Fell, its upper reaches wreathed in fog. The whole vale appeared shrouded in a spectral blue haze, as if it had been submerged beneath the ocean; a thin drizzle began to descend, adding to the sense that we were underwater. As we walked further, I squinted at the approaching hillside, trying to make out the shapes of trees through the milky translucence of the air. Without Lee and Dave knowing the way, I'd have been hopelessly lost.

Then, just as we started to ascend the fell, the cloud momentarily lifted, and I could see Young Wood looming on the vertiginous skyline. What I'd taken to be stony cliffs was, in fact, a sea of grey branches, now outlined against the horizon. With the wood in our sights, we began to climb; gradually at first, and then steeper and steeper, stumbling over scree made slippery by the rain. 'This slate's like glass,' complained Dave, and it was true: in places we had to pull ourselves up by holding on to the straggly outcrops of heather.

Young Wood is nothing if not remote. At its upper extremity, it's a dizzying 485 metres above sea level; half the height of Scafell Pike, England's highest mountain. It exceeds the altitudinal limit of every other upland wood in England: one study describes it as a 'woodland beyond the edge'. And though Young Wood is small – barely two hectares in size – it more than meets the 0.5 hectare threshold to be included on official maps of woodland.[1] But cartographers continue to overlook it, its location marked only by a cluster of tightly bunched contour lines.

Despite this lacuna, Young Wood in the past twenty years has increasingly attracted the attentions of ecologists, who have tried to protect it from overgrazing by sheep through erecting an exclosure fence. As a result, getting to Young Wood is becoming ever more difficult: after navigating the scree, we had to fight our way through a tangle of resurgent gorse that's grown up over the last decade.

But the journey was well worth it. Panting from the climb, we were rewarded by the spectacle of Young Wood as its outline coalesced out of the mist. A huddle of wizened corkscrew oaks hove into view, moss riding up their trunks and coating their branches like fur pelts. The air on the fell, thick as butter, beaded the boughs with condensation. Teal-coloured *Cladonia*, a lichen with tiny antlers, formed forests in miniature on the boles of the trees. The oaks themselves were hardly much bigger: we bowed our heads to pass beneath them. I was astonished to find that some of the oak canopies were only a foot or two above the ground, spreadeagled horizontally against the wind.

Young Wood is only tiny now, but it surely offers a glimpse of how most of the Lake District would once have looked. 'Look at the state of that hillside,' said Lee, gesturing at the far fell. I looked, listening to the distant hiss of the beck in the valley far below, my eye caught by the patches of snow glinting on its upper reaches. They were the only features of note in an otherwise billiard-smooth expanse of grass, picked clean by sheep. Yet Young Wood likely once stretched all the way down this valley, cleaving to both sides. Centuries of deforestation and overgrazing have left the wood as the area's sole remaining outpost of temperate rainforest, too imperceptible even to be noticed by the Ordnance Survey.

Remote, small, fragile and unmapped: Young Wood encapsulates in miniature the plight of Britain's lost rainforests.

When I first started exploring Britain's rainforests, I was astonished to find that no one had yet produced a definitive map of them. A couple of broadbrush maps of the *global* temperate rainforest biome exist, showing both their scattered distribution – from western Europe to Japan to the Pacific northwest coast of America – and how rare they are: rarer even than tropical rainforest.[2] But though scientists have studied temperate rainforests across the world for nearly a century, the rainforest relicts of Britain remain under-recognised and wreathed in mystery.

'Because very few rainforests remain throughout Europe, they have not received much attention from ecologists,' says Dominick DellaSala, author of the first book-length study of temperate rainforests across the world. 'Therefore, broadly applicable classifications are lacking.'[3] When we went looking for mosses in Glen Coe together, the bryologist Ben Averis – who's explored these habitats for decades – told me: 'There is no published definition or description of British and Irish temperate rainforest.'[4] I searched high and low for an official government map of Britain's rainforests, assuming that this globally significant ecosystem would surely have been fully documented and protected – but found nothing.

What I found instead were intrepid efforts by a handful of pioneering ecologists and writers to raise awareness of these incredible places. The author and campaigner George Monbiot was one of the first to popularise the existence of rainforest fragments in Britain in his landmark book *Feral* in 2013. 'A few miles from where I live,' he wrote, 'I have found what appears to be a tiny remnant of the great Atlantic rainforest, a pocket of canopied jungle.'[5]

I lived in Machynlleth at the same time as George was writing *Feral*, and went to explore the spot he described in the Llyfnant valley: a hidden jewel of cascading waterfalls, sheer rock faces drenched in moss, and trees outlined with haloes of polypody ferns. Once, swimming in one of the plunge pools at this magical spot, I found a cave on the far side of the waterfall. Standing in the dripping mouth of the cavern, I tested its acoustics, and listened in awe as the whole hillside seemed to reverberate with my shout. At the time, however, I had no idea of the other rainforest remnants that lie scattered like stars across the British countryside.

Another pioneering attempt to connect the dots in this glittering constellation came with the botanist Clifton Bain's 2015 book *The Rainforests of Britain and Ireland*. It's a gorgeous travelogue, a coffee-table book filled with sumptuous photos and illustrations of rainforest sites from up and down the British Isles. But it makes no claims to be comprehensive, instead presenting a selection of the most publicly accessible sites, to represent the broad geographical spread of where rainforests can occur.[6]

Most tantalisingly of all, I was excited to find a 2016 paper by the ecologist Dr Christopher Ellis, of the Royal Botanic Garden in Edinburgh, which sought to map Britain's temperate rainforest 'zone' – the parts of the country with the right climate to support rainforest.[7] But whilst Ellis's brilliant work takes us much further forward – and I'll be digging into what it shows later in this chapter – its purpose is to define the broad areas in which rainforest *can* occur, rather than pin down precisely what's left.

I wanted to know more, and felt sure that there was more to be done to piece together a more complete map of these rainforest fragments. To me, this wasn't just a matter of curiosity; there was also a more pressing reason to do so. Without knowing where Britain's remaining rainforests actually are, how can we expect landowners and the government to protect and restore them?

So I decided to ask people for help. I started a blog, Lost Rainforests of Britain, set up a Google Map, and shared it on Twitter with a simple request: can you help me map our lost rainforests? To help guide people in the right direction, I suggested a simple criterion: when you're out for a walk in the woods, look for signs of epiphytes – plants growing on other plants. I chose the easily recognisable polypody fern as the clue that rainforest detectives should look out for as a sign that they might be standing in temperate rainforest.

The scale of response took me by surprise. I was deluged with submissions. My Twitter feed filled up with photos of gorgeous wet woodlands that people had visited: fern-lined branches, moss-covered trunks, misty and atmospheric forests. The Google Form I'd created to receive details of potential rainforest locations exploded with hundreds of entries. Email after email poured into my inbox, each one reporting a new discovery, or passing on rumours of a promising location that I should visit.

'Have you looked in the wooded combes of north Devon? They're magical,' read one email. 'Some wonderful rainforest fragments in the various woods around Burrator on Dartmoor!' said another. 'There's little pockets of this all over Cornwall. Been exploring them most of my life.' 'What an exciting and wonderful project!' 'These have always been some of my favourite places to be, I had no idea they count as rainforests.'

Some of the folks who contacted me were biology students, like Katie O'Connor, who told me about a video she was making looking for potential rainforest sites in the Peak District. She was moved to do so, she said, because of 'the importance of conserving these magical pockets of rare beauty'. Others were professional photographers, like Joe Marshall, who lives in west Cornwall and loves exploring the wet woods that line the deep Cornish estuaries. 'I'll be photographing trees for the rest of my life, I reckon,' he said. With infectious enthusiasm, Joe told me about the ancient rainforests of the Helford River, a rare example of an oakwood that grows right next to the sea; high tides leave seaweed hanging in the branches.[8]

What I loved most about the stories people told me was that they spoke to a wider reawakening of public interest in nature. My project seemed to be tapping into something deeper: a re-enchantment with the natural world; a growing awareness of what we've lost; a stirring of dormant memories about landscapes long forgotten. One such recollection came from Naomi, who told me that my map had reminded her of a wood in Wales she'd visited years ago as a child, and now wanted to track down again. 'The trees were low and shapely, emerging out of boulder-strewn ground smothered in luminous moss,' she recalled.

The hundreds of submissions I received told me something else, too, about Britain's remaining temperate rainforests. Plotted together on a map, these potential sites were more numerous than I'd first thought, but each looked tiny in size: small pockets of habitat scattered across the landscape, lurking in wait. That would make detailed mapping of them harder, but all the more rewarding.

At the same time, I also knew that there would be a limit to what citizen science could achieve. Not all the submissions I was sent turned out to be temperate rainforest: some people, quite understandably, mistook climbing plants like ivy (whose roots remain anchored in the soil) for epiphytes (which are anchored only on the tree itself). I also started to fret that I'd made a poor choice of indicator plant. Whilst Britain's temperate rainforests certainly abound in polypody ferns, better clues to the habitat's existence come from the presence of various species of lichens and bryophytes – but these are harder to identify for the layperson, myself very much included. My aim had always been to start people off with an easily recognisable species, and then encourage them to dive deeper using lichen and moss identification guides, such as those published by the charity Plantlife.

Then I started getting sent photos of mossy woodlands from places far outside those which ecologists like Chris Ellis reckon to be the 'rainforest zone'. At first, I was plagued by guilt. Had I sent people off on a fruitless search? Before long, the sense of unease turned to puzzlement. Were these seemingly errant sightings of eastern wet woodland, from Somerset to Sussex, actually a clue that rainforest relics are not wholly confined to the rainy west of Britain? Many of these submissions seemed to be coming from woods in deep valleys and sheltered gorges. Might these places harbour microclimates allowing rainforest to survive?

I had to know more. I decided that the crowdsourced map would be only the first phase in attempting to piece together a more complete and detailed picture of Britain's lost rainforests. I wanted to better understand what was going on here, and delve deeper into what defines a temperate rainforest. To do that, I resolved to undertake a crash-course in basic ecology. My desk started to disappear under books on subjects like climate science, plant identification and woodland conservation. I rang up climatologists, had Zoom calls with botanists and went for walks with lichenologists. Data scientists at government bodies became wearily used to my incessant requests for digital maps and spreadsheets. All the while, I was visiting dozens of rainforest sites around Britain, trying to build up a picture from experience at the same time as assembling one using data. Arthur Conan Doyle's Professor Challenger explored the Amazon armed with compass and machete: my rather less macho voyage of discovery required a botanical hand lens, and a geekish knowledge of digital mapping software.

Where to begin on this journey? To try to draw a map of Britain's remaining rainforests, I've split the process into three parts.

First, we need to look at some climate science, and understand the sorts of climatic conditions in which temperate rainforest can thrive. Second, we have to look at the woodlands growing in these areas, and figure out which might be remnants of rainforest, versus more recent woods that people have planted. And third, we need to factor in the records that botanists have gathered of the species that make our rainforests special – the lichens, mosses, fungi and other plants that inhabit them.

So let's get started by talking about a subject that all Brits are very used to discussing: how much it rains.

* * *

How much rain does a forest need before it's called a rainforest? The ecologist Paul Alaback was one of the first to try to answer this question. In a 1991 paper on rainforests in the Americas, he suggested that the world's temperate rainforest zone could be defined using a number of factors, the first of which was having over 1,400 millimetres of rain annually.[9]

Fourteen hundred millimetres is a *lot* of rain: nearly a metre and a half falling on one area every year! London gets about half that amount: some 722 millimetres of rainfall in an average year, going by the Met Office records for the weather monitoring station at Hampstead in north London. Manchester, despite its reputation for dismal weather, isn't actually much wetter: the weather station at Woodford on its outskirts registers an average of 868 millimetres of rain in a year.[10] But as you head further west in Britain, you soon notice it getting more and more rainy: at least, that was my experience of moving from London to Devon. And the precipitation stats back up this impression. The weather monitoring station at Yarner Wood on the edge of Dartmoor, for example, recorded an average of 1,438 millimetres of annual rainfall over the period 1991–2020 – meeting Alaback's threshold. It just so happens that Yarner Wood is also a rather nice example of temperate rainforest.

But the sheer amount of rain that falls on an area over a year isn't the only factor that matters here. For a rainforest to thrive, it's no good if all that rain gets dumped in one go, only for it to be bone-dry for the rest of the year: the moisture-loving lichens and mosses would shrivel up and die. So Alaback's definition also included an additional proviso: 10 per cent or more of that 1,400 millimetre rainfall threshold had to occur during the summer months. That would imply that an area remains relatively rainy and wet throughout the year, even at its driest moments.

Furthermore, Alaback said, rainfall is only one part of the climatic equation for temperate rainforest: temperature is crucial, too. After all, if an area is very rainy but also very hot, you tend to get *tropical* rainforest – the humid forests of the Amazon, the Congo basin and Indonesia, all sandwiched between the Tropics of Cancer and Capricorn. But *temperate* rainforest, as the name suggests, occurs at middle latitudes where temperatures are mild – western Europe, British Columbia and Japan in the northern hemisphere; Chile, New Zealand and Tasmania in the antipodes. Alaback proposed that regions suitable for temperate rainforest would have 'cool, frequently overcast summers', with July temperatures averaging less than 16°C. If that reminds you of a summer staycation, you may well have been holidaying in one of Britain's temperate rainforest areas.

Until recently, no one had tried using Alaback's definition of temperate rainforest to produce a detailed map of this zone in Britain. Then in 2016, the ecologist Dr Chris Ellis wrote a groundbreaking paper doing so. He plotted where in Britain possesses the 'bioclimatic conditions' for temperate rainforest – more than 1,400 millimetres of rainfall in a year, 10 per cent of that in the summer months, and average July temperatures less than 16°C – using the Met Office's latest data. His resulting map showed that an astonishing 20 per cent of Britain lay within the temperate rainforest zone.[11]

After reading Chris's paper, I got in touch with him requesting a few more details, and he kindly shared with me the digital dataset underpinning his findings – known as a geographic information system (or GIS) map. A GIS map is often more useful than a static map, because you can load it into mapping software, combine it with other datasets, and use it to do calculations. The map that Chris sent me comprised a set of grid squares covering Britain, divided into blocks that met the threshold for temperate rainforest (green) and those that didn't (grey). But, Chris warned, there would be a limit to what the map could tell me. 'I wouldn't give too much attention to the precise position of the boundaries,' he said. 'The calculated distribution would change slightly, depending on the climate data one uses to make the calculation.'[12]

Indeed, the more I looked at the map, the less certain I was that its boundaries were in quite the right place. Ben Averis agreed. 'The temperate rainforest zone in Ellis's map extends eastwards into colder parts of the central and eastern Scottish Highlands, where the woods are floristically very different from the western temperate rainforests,' he told me.[13] To my mind, it also excluded various sites in the southwest of England where patches of temperate rainforest thrive. But then Paul Alaback's original definition of temperate rainforest was intended primarily for use in the Americas: could we even be sure it was wholly applicable to Britain?

There are, after all, other ways to combine data about temperature and rainfall, and other places where one might draw the line delineating where rainforest can exist. One such alternative measure is the 'oceanicity' of the climate. It's something that ecologists in Britain have been studying long before definitions of temperate rainforest existed. Arthur Tansley, the grandfather of British ecology, described in the 1940s how the 'climate along the western coastlands of Europe is what is called *oceanic*, and in this case *Atlantic*, because the dominant factor is the prevalence of south-westerly winds from the Atlantic Ocean, reducing both winter cold and summer heat and increasing the moisture of the air'.[14] Within Britain, Tansley wrote, 'the prevalent moisture-laden west winds have a much greater effect on the climate of the western parts of the country'.[15]

One way of measuring oceanicity is to use what climate scientists call Amann's 'index of hygrothermy', named after the Swiss botanist who invented it, Jean Jules Amann. 'Hygrothermy' is a scary-sounding word but with a simple meaning: *hygro* comes from the Greek for 'moist', and *therm* means 'heat'. Like Alaback's definition, an index of hygrothermy is an equation combining average annual rainfall and temperature, but it also adds in to the mix a temperature *range* for the warmest and coldest months – rather than just specifying cool summers. This helps define the 'Goldilocks zone' in which the climate can be called oceanic: not too hot, and not too cold, but just right.[16]

Having an index allows us to assign a numerical value to an area – to score its oceanicity and plot it on a map, rather than simply describe it. Various botanists have used Amann's index to help explain the distribution of mosses, liverworts and lichens across Britain's western seaboard. One early study, from 1960, found that 'several features of the Atlantic climate contribute to the luxuriant growth of bryophytes, among them the high rainfall and the mild climate without extremes of temperature'.[17] At that time, however, fewer weather monitoring stations existed to collect detailed data on rainfall and temperature, and they were particularly lacking in western Britain.[18] But, since then, the Met Office has amassed much more data, allowing the creation of more detailed maps showing Britain's oceanic zone.

Rereading Chris Ellis's paper, I realised that he had also created a map of oceanicity for comparison, making use of more recent Met Office data and putting it through Amann's index. In his view, an index score of more than 100 puts a grid square in the oceanic zone – a region covering a very slightly larger area of Britain than the rainforest zone defined by Alaback. But the map this produces cuts out the colder parts of the Scottish Highlands which looked anomalous in Chris's other map, and extends over more of the Westcountry. The index score also doesn't stop at 100: it carries on up to a top score of 338 in some of the very rainiest parts of western Scotland. This led Chris to propose that scores over 150 should be considered 'hyper-oceanic': the wettest of the wet, covering about 8 per cent of Britain, and concentrated in western Scotland, parts of Wales, the Lake District and Dartmoor. It suggests we might search for temperate rainforests over a slightly broader oceanic zone, but we can expect to find the most luxurious examples where the climate is hyper-oceanic.

Despite my desire to draw lines on a map, I started to realise I wasn't thinking very ecologically. It can be hard to discern hard and fast boundaries in nature, with habitats instead blurring into one another as the climatic conditions change. As Ben Averis told me, 'rather than try to describe a clear "black & white" rainforest / non-rainforest threshold, it is perhaps more realistic to recognise a continuum of variation'.[19] Chris Ellis agreed: 'There is fuzziness around the edges.' It's a question of scale, he suggested: 'If you step back and look at Britain from a distance, then you see the broad outline of the rainforest, as in these GIS maps – the core areas. But, if you step into a woodland, it's always possible that you might see rainforest communities outside of this core area, given the right microclimatic conditions.'

Chris had raised a very interesting point: what about microclimates? A microclimate can be created where the topography of a place shelters it from the weather. Imagine a deep ravine, in which the rocky cliffs cast shade, keeping the area cool and damp, and shelter plants from the wind, reducing transpiration. A cascading stream at the bottom of this gorge also adds to the place's humidity. Such features would alter the climate of the locality, making it more conducive to rainforest, even if it were in a part of the country that's generally drier and sunnier. I thought back to the photos that people had sent me of fern-filled gorges and moss-covered gullies from Somerset to Sussex. These sites were way outside Britain's western oceanic zone, whichever way you looked at it. But could they be evidence of rainforests surviving further to the east as a result of sheltered microclimates?

Some ecologists have certainly found examples of this. The lichens expert April Windle, for example, told me about a survey she carried out for the Bunloit Rewilding Project in the central Highlands of Scotland. Here she found 'ravines where you have microclimatic conditions that foster rainforest communities, despite this being slightly outside of the western rainforest zone'.

Another potential rainforest microclimate is Lud's Church in the Peak District, which is thought by literary scholars to be the real-world location of the Green Chapel in the Arthurian legend of Sir Gawain and the Green Knight.[20] The 'church' is, in fact, a deep natural chasm, buried in a forest, whose walls glow green from their thick coating of ferns, mosses and wood-rush. Though Lud's Church lies outside Britain's oceanic zone, the microclimate effects within its narrow confines are said to be striking, causing it to be much cooler and damper than the surrounding area, even on hot summer days. If it is true that Lud's Church provided the inspiration for the Green Chapel, it raises the tantalising possibility that the legend of the Green Knight sprang from one of Britain's rainforest microclimates: a place that stays green all year round.

Standard rainfall measurements also fail to take into account phenomena like fog and dew, which deposit tiny water droplets that seldom register on rain gauges. (The technical term for this is 'occult precipitation', meaning 'concealed' – though I like how it also makes fog sound like the weather of witches.) On the Pacific northwest coast of America, some temperate rain-forests are sustained by a belt of fog, of the sort that creeps its way into the pages of Raymond Chandler novels. The Atlantic sea-fogs blowing in over the north coasts of Cornwall and Devon could also help explain why they, too, are peppered with tiny patches of rainforest, like the shrunken oaks of the Dizzard, our very own elfin cloud forest.[21]

Places like Lud's Church and the Dizzard are few and far between; but a potentially much bigger collection of rainforest microclimates lies in the so-called 'ghyll woodlands' of southeast England. A 'ghyll' (or 'gill') is a steep-sided ravine, and there are perhaps a thousand of them etched into the Wealden hills of Sussex and Kent. I was alerted to their existence by the botanist Dave Bangs, an indefatigable campaigner for public access to nature, whom I first met on a mass trespass on the

South Downs. Dave's 'guerrilla botany missions' through the Sussex countryside in search of rare plants have spawned several encyclopaedic books on the subject and made him a legendary figure in his home town of Brighton.[22] One day, I received an email from Dave out of the blue. 'Your maps of rainforest omit the Wealden basin!' he exclaimed.

At first, I tried to ignore Dave's emails: my head was already spinning from trying to understand the intricacies of climate science, and I couldn't face further complexity. Then I became sceptical: rainfall over the Weald is not much greater than it is in London, and the climate is closer to that of continental Europe. But Dave is nothing if not persistent, and he posted me several photocopied studies of ghyll woodlands as evidence. One of them, by the renowned lichenologist Francis Rose, argued that the ghylls are a stronghold of 'oceanic or Atlantic species' of mosses and ferns, and that they have 'more in common with southwest England . . . and with Wales' than with the eastern half of Britain.[23]

In fact, Tunbridge filmy fern – an oceanic fern species that I've found plenty of in the Westcountry – is named after its unlikely sounding original discovery at Royal Tunbridge Wells, in the heart of the Weald. Dave Bangs describes it as living there 'in exile, hundreds of miles away from its nearest temperate rainforest and mountain hangouts on the western seaboard'. The reason for this extraordinary survival likely lies in the sheltered microclimates generated by the deep ghylls – perhaps aided by the porous, water-retaining rock formations of the Weald.

Some conservationists prefer to see ghyll woodland as a distinct habitat in its own right, unique in the UK.[24] But Dave Bangs reckons that it is in fact a 'disjunct population' of rainforest, separated from the bulk of the Atlantic oakwoods following centuries of deforestation. *Weald*, after all, is an Anglo-Saxon word for 'wood'. This landscape was once clothed by the lost forest of Andredsweald, which records suggest still covered a staggering 70 per cent of the Wealden hills by the time of the Domesday Book.[25] Much has since been lost to industry and the plough, but the ghyll woodlands persist as relics.

Though many microclimates are created by topography – rocks and hills – what I find really fascinating is how the presence of *woodland itself* can generate a microclimate. After all, trees can provide shade and shelter from the wind, just like overhanging rocks, allowing the interiors of woods to remain cooler and damper than their exteriors. One recent study has found that temperatures beneath the canopy of European woodlands are 2°C cooler in summer – and 2°C warmer in winter – than in surrounding areas of open countryside.[26] A microclimate even appears to be operating within tiny Wistman's Wood on Dartmoor. Two naturalists, Leslie Harvey and Douglas St Leger-Gordon, recognised as long ago as the 1950s the 'striking contrast' between the plants thriving inside Wistman's Wood and those of the neighbouring moorland. They saw this as 'good evidence of the powerful influence of the new microclimate created once the wood has become established'.[27] Woods, in other words, can themselves generate climatic conditions more conducive to the damp-loving mosses and lichens that characterise rainforests.

This raises another exciting possibility. By allowing woodland cover to regenerate, could we actually see an expansion of the temperate rainforest zone? The microclimate effects are unlikely to be sufficient to create rainforest conditions in very dry and exposed parts of the east of Britain. But they might help tip the balance where the topography is already conducive – by letting canopy cover spread up steep valleys and along the edges of streams, for example. Reading about microclimates, I'm reminded of James Lovelock's Gaia hypothesis: the idea that the biosphere acts as a self-regulating organism, generating conditions to sustain life.

But humanity, of course, has thrown a spanner in the works of this delicately balanced system. The climate crisis, left unchecked, threatens to dry out our remaining fragments of temperate rainforest. As the world heats up, Britain is forecast to experience wetter winters, but also drier summers. Ending the burning of fossil fuels and radically slashing greenhouse gas emissions is a prerequisite for their survival – and for our own. Yet adaptation is also now essential. We have to let our rainforests grow more resilient to the pressures of climate change, and that means giving them the space in which to spread once more.

So much for the role of rain in generating temperate rainforests. But what about the sorts of woods that comprise our rainforest habitats? Can existing maps of woodland help us locate lost rainforests?

I used to think that there were just two woodland types in Britain: deciduous (or broadleaved) and coniferous (or evergreen). Sure, I knew that there were lots of different species of tree: oak, birch, holly, ash, and so on. But I'll confess that, when going on woodland walks, I didn't really think much about the overall composition of a wood. It didn't occur to me how a wood might be defined by the main species present in it – and how this could lead to a much wider set of woodland categories than simply those that lost their leaves in winter and those that kept them.

That was before I was introduced in my thirties to the woodland wisdom of three veteran sages of the forest: George Peterken, John Rodwell and Oliver Rackham. These legendary woodland ecologists are perhaps the nearest thing modern Britain has to the druids of old – not in the blood-and-mistletoe sense, but in the true meaning of the word: *oak-seers*, those with deep knowledge of trees. (The analogy gets stronger when you know that one of them had a very fine white beard, and another is a priest.) During the 1970s and 80s, these three wise men set about cataloguing Britain's woodlands, from the largest forest to the smallest coppice. Their decades of work, for the government's nature watchdog and at the universities of Lancaster and Cambridge, have transformed our understanding of Britain's woods, painting a far more detailed picture of how the country's woodlands can be classified botanically, geographically and chronologically.

'Any attempt to describe British woodland types is bedevilled by the absence of a stable, well-known and widely accepted classification,' wrote the first of our wise men, George Peterken, in 1981. Up until that point, ecologists had continued to make do with a broadbrush system established by Arthur Tansley half a century before, which simply divvied up woods according to their most dominant species: oakwoods, ashwoods, beechwoods and so on. But as Peterken pointed out, 'most woods consist of mixtures'.[28] He set about devising a new classification based on these mixes of tree species, identifying the typical combinations that grow alongside one another in the same wood.

Peterken proposed splitting up British woodlands into twelve main groups, with a total of fifty-eight subdivisions. This level of gradation might sound mind-boggling to you and me, but what Peterken was trying to do was better reflect the dizzying complexity of the natural world, based on his own observations. Simply painting in primary colours, he reasoned, did no justice to the subtle palette of shades existing in nature. In place of 'oakwoods', for example, Peterken advanced that they nearly all comprised a blend of species, with either birch or hazel providing the understorey to the stands of oak. He also proposed differentiating between woods comprising one of the two main oak species – sessile (*Quercus petraea*) or pendunculate (*Quercus robur*) – and between upland and lowland oakwoods.[29]

Getting to grips with the sheer diversity of our woodlands is vital for better understanding the species that make up Britain's

The 'lonely wood of Wistman' has given rise to many legends and tales of gothic horror. The 'Wisht hounds' said to haunt it likely inspired Sir Arthur Conan Doyle's Sherlock Holmes novel, *The Hound of the Baskervilles*.

Filmy ferns (such as *Hymenophyllum tunbrigense*) can be excellent indicators of rainforest habitat. They thrive in damp gorges and near streams, glistening like seaweed in a rockpool.

Ochrolechia tartarea, a 'cudbear' lichen, has fruiting bodies that look like jam tarts. Historically, it was used to make dyes.

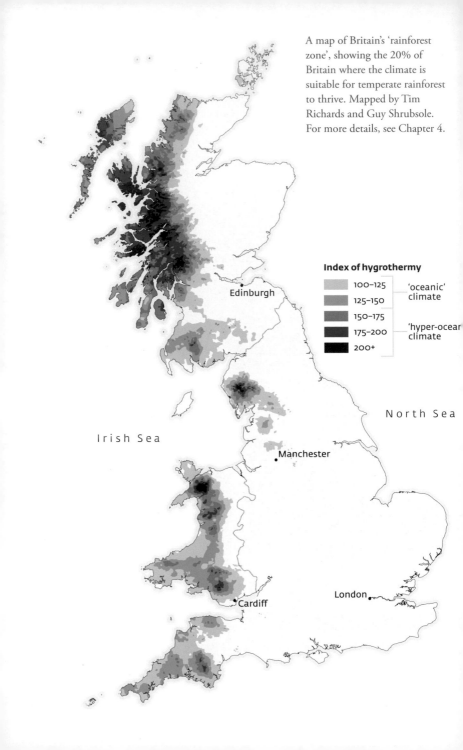

A map of Britain's 'rainforest zone', showing the 20% of Britain where the climate is suitable for temperate rainforest to thrive. Mapped by Tim Richards and Guy Shrubsole. For more details, see Chapter 4.

Index of hygrothermy

100–125   'oceanic' climate
125–150
150–175   'hyper-ocean[ic]' climate
175–200
200+

Edinburgh

North Sea

Irish Sea

Manchester

Cardiff

London

A map of Britain's surviving temperate rainforest fragments, covering less than 1% of the country. Mapped by Tim Richards and Guy Shrubsole. For more details, see Chapter 4.

North Sea

Irish Sea

Edinburgh

Manchester

Cardiff

London

German biologist Ernst Haeckel's print of tree lungwort, *Lobaria pulmonaria* (centre), alongside other lichens, 1899.

Lobaria scrobiculata, or 'Lob scrob' for short, is the gunmetal-blue cousin of tree lungwort.

There are various species of *Sticta* lichens, many of which have the dubious distinction of smelling like rotting fish when wet. This photo shows *Sticta sylvatica,* spotted in Ceunant Llennyrch, north Wales.

'String-of-sausages' lichen (*Usnea articulata*) hangs like an eggshell-green necklace off branches in West Country rainforests, sometimes festooning entire trees.

Royal fern (*Osmunda regalis*), the crowning glory of Britain's rainforest ferns, whose fronds can grow up to three metres tall. This species suffered from over-collection by Victorian 'fern maniacs'.

Degelia atlantica, a lichen of the very wettest parts of Britain's hyper-oceanic zone, resembles a cluster of scallop shells, forming a velvety grey mat over branches.

What really marks out a temperate rainforest is the wealth of epiphytic plants growing on the trees. The branch of this oak tree, from the Dart Valley in Devon, has practically disappeared under the weight of mosses and ferns growing on it. The rich soils that form in the canopies of our temperate rainforests may be a substantial carbon sink, making them vital allies in tackling the climate crisis.

Polypody fern *(Polypodium vulgare)*, whose name means 'many-footed', creeps along branches by sending out horizontal roots, lending Britain's rainforest trees a green halo.

The Dizzard in Cornwall, an ancient rainforest of gnarled and shrunken oaks, is thought to be thousands of years old. In places, the trees are so small, you can raise your head above the forest canopy.

The ancient temperate rainforests at Ceunant Llennyrch and Coed Felenrhyd in north Wales are not only ecologically important, but also culturally significant: they are the site of a legendary battle in *The Mabinogion,* a book of Welsh myths.

A visit to a rainforest feels to me like entering a green cathedral. Moss-encrusted trees line this footpath near Relubbus, Cornwall.

temperate rainforests. Whilst it's true that many of our rain-forests are dominated by oak – sometimes as a result of historic selection for oak coppice – they also contain multiple species, including hazel, ash, birch and holly. And though it would be nice to be able to generalise and say that most of our rainforests comprise upland sessile oaks, it wouldn't be true – Dartmoor's three upland oakwoods are all pendunculate oaks, for example. Such a generalisation breaks down still more the further north you go, with many temperate rainforests in the Lake District being dominated by ash, and with Scotland playing host to extraordinary rainforests of Atlantic hazel and Caledonian pine. So, could Peterken's system help me map Britain's rainforests, I wondered?

Unfortunately, systems of woodland classification seem to be like London buses: you wait ages for one to come along, only for several to turn up all at once. At the same time that Peterken had proposed his typology, an even more complex system was being devised: the National Vegetation Classification, or NVC. The NVC was the brainchild of the second of our wise men, John Rodwell, an ecologist who is also an Anglican priest in the diocese of Blackburn. Rodwell – working with a team of other ecologists and botanists – intended the NVC to be a cataloguing system for *all* plant communities in Britain, not just woodlands. This vast undertaking, unsurprisingly, took fifteen years to complete. When it was finally published in 1991, it quickly superseded Peterken's categorisation, and was taken up by government environmental quangos as the 'official' system for woodland classification.[30]

Rodwell's NVC differed from Peterken's approach by adding in ground vegetation: the species of plants that grow on woodland floors, beneath the trees making up the canopy and the bushes and shrubs comprising the understorey. Woodland and scrub habitats were divided up into twenty-five types, each with an identification code: 'W1' woodland, for example, refers to grey willow woods with a ground covering of marsh-bedstraw. The inclusion of ground flora meant all the layers of a wood were being represented, but also perhaps made the system harder for the general public to get their heads around. W17 woodland, for instance, is known formally as '*Quercus petraea–Betula pubescens–Dicranum majus* woodland': or, in plain English, a wood dominated by sessile oak, downy birch and greater fork-moss.

As I tried to grapple with these increasingly labyrinthine systems, a question was nagging at the back of my head. Where, in all of these classifications, might British temperate rainforest fit? Whilst both Peterken and Rodwell were clearly very familiar with the wet woodlands of western Britain, nowhere in their tomes did the word 'rainforest' appear. They spoke instead of 'western oakwoods in oceanic climates', or woodlands containing 'Atlantic species' of mosses and liverworts.[31] John Rodwell's descriptions of W17 woods came closest to the temperate rainforests I knew and loved, speaking of how 'oak forms a very dwarfed canopy' in upland woods like Wistman's and Keskadale; how 'epiphytic lichens can also be very prominent' in such woods, with 'dripping mats of bryophytes'; and how 'in sheltered ravines . . . [a] fern element can be very extensive and luxuriant'.

But, I found, the concept of 'temperate rainforest' seemed to have fallen through the gaps in these different systems of classification. The NVC, for example, tends to ignore most epiphytes, focusing instead on the mosses that grow on the forest floor; and it has little to say about lichens.[32] And perhaps, back in the 1980s, 'rainforest' just seemed too exotic and alien a concept to apply to Britain. Environmentalists at that time were only just starting to raise the alarm about the destruction of the tropical rainforests, whose pristine character seemed a million miles away from the heavily managed and fragmented woods of little old Blighty.

I wanted to see if I could salvage something from these undoubtedly brilliant endeavours to help build a map of British rainforests. By chance, I realised that I'd been exchanging emails with John Rodwell for the best part of a year without realising who he was. He and I had been working together to persuade the Church of England to reforest its huge landholdings. 'Wait, you're *the* John Rodwell, who wrote the NVC?' I fanboyed. 'Steady on!' replied John, adding modestly, 'The project was very much a team effort.' Could the NVC be used to help map temperate rainforest? I asked.

John sent me in the direction of a dataset of woods broken down by NVC type, published on the website of a government environmental body. Thinking this could allow me to make a map, I downloaded the data but found, with mounting disappointment, that it was a mess. Many of the point locations were wrong or came with only rough grid references, dating from a time before GPS enabled pinpoint accuracy. This was a big problem, because it meant I couldn't easily merge the NVC data with other maps showing the boundaries of woods: the fuzzy location data meant some woods would be left out, and other woods would be included in error.

What I really needed was more accurate cartography: something called a 'polygon map', showing not simply a point location, but the exact boundaries of a woodland broken down by NVC type. But although I searched high and low, such maps simply don't exist for most of the country. I found some for Scotland, and Richard Knott, the ecologist for Dartmoor National Park Authority, kindly sent me a map delineating NVC types across Dartmoor.[33] But when I asked the other National Park Authorities covering western Britain, none of them had any such maps; and nor did the UK government's green watchdog, Natural England. I asked John Rodwell about this gap in the ecological record. 'Vegetation maps of many woodlands exist but, across England, this was done piecemeal for different projects,' he sighed. 'There has been lamentable after-care of the NVC in general, with no single UK vegetation database, no standardisation of reporting, no bibliography of reports.'[34] I got the sense that this huge labour of love had only just become established before official bodies abruptly gave up on it – either as a result of shifting scientific opinion, or cuts in public funding.

I felt like I'd hit a dead end. Mapping the climate zone in which rainforests thrive had turned out to be far from straight-forward, and now it seemed that even mapping the forests themselves might prove impossible. But then came salvation from the third of our wise men of the woods, Oliver Rackham.

Rackham died in 2015, a colossal loss to the world of conservation. Whilst I couldn't ask him about Britain's rainforests in person, I delved deep into Rackham's wonderful books and talked to conservationist friends who had met him. 'You'd have loved him,' said Kevin Cox, chair of the RSPB and fellow Devon resident. 'To see a wood through his eyes was fascinating, utterly gripping. As a character, he was straight out of Tolkien.' Old photos of Rackham show him wearing green tweed the colour of Frodo's cloak, and sporting a beard as white as Gandalf's.

Kevin suggested I read some of Rackham's field notebooks, many of which have been digitised and put online by Cambridge University, where he was an academic. Written in pencil and faded ink, their well-thumbed pages read like prayer books to the woods in which Rackham worshipped. They're saturated with lovingly detailed accounts of the woods he walked in: lists of each and every species he encountered, remarks upon rarities, observations about each woodland's history, the margins of their pages filled with sketches and hand-drawn maps. Much of Rackham's work focused on the relatively dry woods of Essex, but he also fell in love with the western oakwoods of the Helford River in Cornwall. His notebooks from these visits come closest to describing rainforest, noting trees that are 'crooked and branchy . . . full of bryophytes, with polypody growing on limbs and even high in the crown'.[35]

But what I found most fascinating about Rackham was his pioneering emphasis on using history as a guide to different types of woodland. Classifying ecosystems by their mix of species is one way of doing it (and Rackham had his own preferred system for this, too). Yet where other woodland ecologists cared about *space* – the geographic distribution of different species – Rackham cared just as much about *time*. His books – from *Trees and Woodland in the British Landscape* (1976), to his magnum opus, *Ancient Woodland* (1980) – opened people's eyes to the importance not just of woods, but how old they were. Rackham could read a wood like others read a historical text: in its veteran trees, coppice stools and boundaries, he discerned the imprint of centuries of human interaction.

The more ancient the woodland, Rackham showed, the more time it had to develop into something richer and more diverse – becoming a home for those species of plants and fungi which take a long time to become established. Species, of course, like the many rare lichens which characterise our rainforests, whose slow reproductive cycles mean they only spread slowly from branch to branch. Could a map of *ancient* woodland, I wondered, be a good proxy for where some of our rainforests survive?

Fortuitously, as a result of Rackham's tireless work, maps of ancient woodland across Britain do indeed exist. Following the publication of *Ancient Woodland*, and the subsequent surge of public concern for the plight of our ancient woods – many of which had been felled in the decades after the Second World War – the authorities finally began to show an interest in them. In 1981, a team of ecologists at the UK government's Nature Conservancy Council – including the first of our wise men of the woods, George Peterken – began compiling an Ancient Woodland Inventory for the country. By using old maps, aerial photos, tithe surveys and historical records, they painstakingly pieced together a map of Britain's woodland history – a 'formidable task' that Rackham had originally 'thought impossible'.[36] 'Ancient woodland' is today defined as any wood that's been in continuous existence since the year 1600 in England and Wales, or since 1750 in Scotland. Many of these woods will also have histories stretching back hundreds if not thousands of years earlier.

Focusing on ancient woodland – rather than trying to differentiate by the dominant tree species or mix of species – has other benefits, too. Although many of our temperate rainforests are dominated by oak, many also comprise ash, birch, hazel and other species. What matters is whether they're growing in areas where the climatic conditions are right for them to support epiphytes. Continuity of woodland, rather than the precise NVC type, is ultimately a bigger factor. And published maps of ancient woodland, whilst still incomplete, are also very detailed – drawing in the precise boundaries of woods, rather than just plotting a set of vaguely located points.

Now I was armed with a better understanding of woodlands and climate science, I felt more equipped to trace the outlines of Britain's rainforest relics.

I teamed up with Tim Richards, a professional GIS mapmaker, and together we set to work mapping rainforests. I started by stitching together maps of ancient woodland in England, Wales and Scotland – since devolution in the late 1990s, each inventory is maintained by separate nature bodies – to create one map of ancient woods across the whole of Britain. Then Tim tracked down the latest Met Office data on rainfall and temperature records, allowing us to update Christopher Ellis's map of the country's oceanic zone and draw it at a higher level of detail. We then used that information to clip our map of ancient woodland, in the same way as one would use a cookie-cutter to cut shapes from pastry – leaving us with a map of ancient woodland in those western parts of Britain where the climate allows rainforest to thrive.

A further breakthrough came with help from Ben Averis, who sent us a list of possible indicator species for temperate rainforests. As we learned in Chapter 3, the presence of certain species and communities of lichens, mosses, liverworts and ferns can act as vital clues to the existence of rainforests. When it comes to botany, Britain is one of the most surveyed countries on earth, but there are still plenty of surprises to be uncovered – both through botanising in the field and through analysing the vast datasets amassed over the decades. A lot of botanical records are unpublished, or kept behind paywalls at cash-strapped local environmental records centres. But fortunately, the nationwide datasets of the British Lichen Society and British Bryological Society have been made freely available online. Tim was therefore able to download records for the species on Ben's list, see where they occur across Britain, and combine them with our map of ancient oceanic woods.

The resulting set of maps, shown in the colour plates, are a first attempt at mapping Britain's remaining rainforests. They're not comprehensive; I hope they can be improved upon; but they're a first go. My excitement at being able to piece together these maps, however, is tempered by gloom.

Measuring the area of potential temperate rainforest left in Britain reveals a tragic statistic. At most, around 333,000 acres remains – about half a per cent of Britain's land surface. A breathtakingly biodiverse habitat that perhaps once covered a fifth of Britain has been smashed into fragments covering less than 1 per cent of the country.

Even this is likely an overestimate. Only a quarter of the ancient woods in Britain's oceanic zone have botanical records for rainforest species of lichens, mosses and liverworts.[37] To put it another way, out of the old-growth woods we've identified as having the right climate for rainforest to thrive, only around 25 per cent of them are known to have rainforest species of lichens and bryophytes present. This could be because many rainforest sites haven't yet been properly surveyed; or it could mean the remaining area of healthy rainforest is even smaller. The Alliance for Scotland's Rainforest has made a similar calculation, suggesting just 74,000 acres of lichen-rich rainforest survive in Scotland, even though the total area of ancient oceanic woods left is larger.[38] So much has been lost; so little remains.

But we didn't just lose our rainforests through some tragic accident. We actively destroyed them. What follows next is the story of how we did so.

5

Myth, Magic and The Mabinogion

The A470 travelling north through Wales did not, at first, appear a promising site for temperate rainforests. The twin dark cubes of Trawsfynydd nuclear power station sulked malevolently on the horizon, whiling away their radioactive half-lives. Llyn Trawsfynydd – the man-made reservoir on whose shores the decommissioned reactors squat – lurked with a glassy blackness.

Slot-eyed sheep picked fretfully at the grass in the surrounding fields. I found it hard not to think of the flocks slaughtered prematurely when the Chernobyl disaster dumped its radioactive cloud over the mountainsides of north Wales in 1986, the year after I was born.

But just beyond the rotting hulk of the power station, I knew, lay a rainforest. Earlier that summer, during a mass trespass on the South Downs, I'd spoken with a film-maker, Emma Crome. Emma told me she lived on the edge of a spectacular fragment of Atlantic oakwood, buried deep in Snowdonia National Park. 'Come visit!' she said. Naturally, I jumped at the chance.

So Louisa and I set off together on another rainforest road trip. We packed our car – a red Vauxhall Viva affectionately nicknamed Zoomer: an ironic reference to how it copes with Devon hills – with our sleeping bags, tent, camping-gas stove and flasks of tea and coffee. Listening to the Catatonia singer Cerys Matthews rolling her Rs helped stave off road rage as we wound our way slowly through holiday-season traffic. Zoomer became increasingly strewn with crumbs from the oatcakes and hula hoops we munched along the way.

We stopped off in Hay-on-Wye to buy books – including a faded copy of *The Mabinogion*, a collection of ancient Welsh legends – and to pick up our friend Jon Moses, who was joining us for the adventure. Our journey north took us past dark conifer plantations and sheep-dotted hillsides, circled by buzzards and red kites. At one point, we watched, wide-eyed, as a stoat scampered across the road in front of us.

The approach to Emma's house was not so much a road as a dirt track, punctuated by gates to stop sheep migrating to neighbouring fields. As we pulled up we were greeted by Emma's dog Merryn, an excitable young spaniel whose front and back legs appeared to move in completely different directions. Emma took us on a tour of her home, an old farmhouse she'd recently moved into with her partner Guy. Both of them were lean from years of rock-climbing, their Rab jackets and polyester fleeces clues to their time spent outdoors mountaineering. Emma's bookshelves were filled with titles like *Peak Limestone*, *Eastern Grit* and *Over the Moors*; an ice axe hung from the wall in their spare room. Although she originally hailed from Devon, moving to the mountains of Snowdonia had held an obvious appeal for Emma.

That evening, Louisa, Jon and I sat with Emma and Guy around their big kitchen table eating delicious veggie curry and poring over Ordnance Survey maps, plotting the following day's adventure: a descent into the ancient rainforest of Ceunant Llennyrch, which lay just over the brow of the hill.

Ceunant Llennyrch is a steep-sided ravine, down which the Prysor River cascades. *Ceunant* is Welsh for 'gorge', whilst *llennyrch* means 'glade'. Together with the neighbouring Coed Felenrhyd – *coed* meaning 'wood', *felenrhyd*, the 'yellow ford' – it forms a National Nature Reserve. The conservation group Celtic Rainforests Wales says that 'the steep banks of the Afon Prysor are thought to have been wooded for thousands of years – possibly since trees first re-colonised Wales after the last Ice Age'.[1]

The next morning, caffeinated and breakfasted, we set off to explore the rainforest. Our route into the gorge involved first passing beneath the huge pipeline that snakes across the landscape, ferrying water from Trawsfynydd reservoir to the hydroelectric station in the valley below. Above us thrummed the insistent churring static of pylons, carrying electricity back up the valley in return. A flock of fiery-tailed redstarts swooped overhead, unperturbed by the back-and-forth of water and electrons.

Jon, a keen fungus-forager, spotted various delicacies along the way: a waxcap growing amongst wild strawberries, a bolete sprouting from a birch. Even he, however, struggled to identify some species.

'What's this one?' asked Emma.

'Ah, that one's what mycologists call an LBM,' said Jon.

'An LBM?'

'A Little Brown Mushroom.'

I've come to know Jon through our campaigning together for greater public access to the countryside. Jon is Welsh, lives in Monmouth and has a way with words that veers swiftly from the academic – he has a doctorate in the politics of phenomenology – to laddish banter. His definition of what counts as a rainforest is pithier than mine: for him, they're 'places where shit grows on shit growing on other shit'. Certain furry lichens 'have a pubic quality' in Jon's eyes. It's true; they do.

In a clearing, we paused to press our hands into the hummocks of sphagnum and star moss erupting from the forest floor. I decided to sink my entire face into one to smell the earthy humus scent: cue embarrassing photos to be used as blackmail material later. Around us sprouted the blood-orange berries of rowan, the lilac shades of heather, young birch saplings, a late-flowering foxglove.

As the fields turned into forestry plantations, we encountered a party of canyoneers, suited, booted and helmeted, clearly pumped up for a day of scrambling. After they had passed, Emma confided to us the reservations she now held about climbing culture, talking wryly about the 'polyamide-wearing elite' who seem more intent on conquering nature rather than experiencing it. As it happened, we were set to do some canyoning of our own, only with less gear.

Before we could start the descent into the gorge, however, we had to traverse the dam at its head. A vast, brutalist buttress, its beige-grey concrete walls hold back thousands of tons of water, stoppering up the valley like a cork. Looking to one side, out over the jet-black surface of the reservoir, we heard the mournful cry of a wading oystercatcher.

'I always stop and wonder what it would have looked like before the dam,' Emma mused. Old maps of the area from the nineteenth century, before Llyn Trawsfynydd was built, show most of the land was a bog, but also reveal a former farmstead with the name Llwyn-derw, Welsh for 'oak-grove'. Another nearby house was called Is-law'r-coed, meaning 'underwood'; either referring to a coppice, or the house beneath the oaken grove. Whilst some of the oaks appear to have survived on isolated islands, the farmsteads and most of the grove were drowned with the construction of the reservoir in the 1920s.

What still survives, however, is revealed by looking over the other side of the dam. We peered over the concrete lip to gaze fretfully at the vertigo-inducing 100-foot drop below, into what appeared to be a primeval valley cloaked in vegetation.[2] Birch and oak clung precipitously to the sides of the cliffs carved by the river far beneath us; heather and bracken seemed to tumble towards the clatter of rocks at the dam's base.

We began our descent. On the way down, I was entranced first by a blue-black beetle, then by a glistening beefsteak fungus growing from a tree, its red flesh bearing a disturbing resemblance to muscle tissue. In winter, when beefsteak fungus starts to decompose, it takes on more of a tea-time quality: to me, it looks like jam scraped over a toasted pikelet.

The further down we went, the moister and more humid it became. Vegetation in the gorge responded accordingly, luxuriating in the abundant wetness. The plant life growing on a fallen log next to a stream signalled, to my delight, that we had truly entered rainforest territory: an exuberance of *Sticta sylvatica*, the fish-smelling lichen, competing for space with an emerald mat of mosses and liverworts.

For a moment, Louisa and I thought we'd lost Emma and Jon: they'd disappeared from sight up ahead, and the wood was quiet. Then an eldritch voice hissed from below. '*MY PRECIOUSSS!*' We looked down and Jon and Emma were hiding in a tiny cave, probably an old adit from former mine-workings, its entrance almost wholly concealed by leafmould and grasses. Jon stared up impishly, his face seemingly suspended in the darkness of the cave as he gleefully burbled his Gollum impression.

We reached the end of the 'official' path taking us to the valley floor, where the Prysor River rushes down to the yellow ford, Y Felenrhyd. The plan now was to climb back up along the gorge, heading towards the dam once more. This part of Ceunant Llennyrch is seldom visited by anyone, requiring much scrambling over rocks and wading through the plunge pools of waterfalls. We ducked beneath a hazel branch weighed down by the quantity of filmy ferns sprouting from its back, and pressed on.

Over rocks greased with moss; over a fallen trunk slippery

with algae; over a still pond, we trod gingerly on stepping-stones in our welly boots. A tiny newt, disguising itself as a stick, peered from a cleft of rock dripping with liverworts and filmy ferns. An oak protruded horizontally from one side of the ravine like an outstretched arm, its bough thick with humus and sphagnum. Now the gorge had become vertical-sided, entire walls of moss studded with woodrush. I looked down into a boulder-filled plunge pool, its depths the blue-green colour of weathered copper. There were no signs that anyone had been here before: no graffiti in the moss, no discarded sandwich wrappers. Small wonder: the going was tough, and we were all knackered.

Something else made the rainforest here feel distant from the human world. As we entered the uppermost reaches of the ravine, we passed a wasps' nest. Suspended high in the branches of an ash, it was seemingly abandoned. But as we turned to head back, the gorge became filled with an incessant droning. We couldn't see the wasps, but we got the message that we were no longer welcome. I felt the eerie sense that this place wasn't inhabited by people but rather by some other force, one that was both alien and also wholly natural.

Yet this rainforest is also steeped in human history. It's not just significant ecologically, but also culturally: it is the site of a legendary battle in *The Mabinogion*. In fact, as I later discovered, rainforests arguably form a foundational part of Welsh culture.

The Mabinogion is a collection of some of the oldest, weirdest myths of Britain: stories of blood and magic, warriors and wizards. It's hard to overstate their centrality to Welsh history, literature and culture.

First written down around a thousand years ago, the tales of *The Mabinogion* undoubtedly have an even older oral prehistory, perhaps dating back to 'the dawn of the Celtic world', to quote one of its translators.[3] The stories comprise a loose compilation of legends rather than a single narrative; the term *The Mabinogion* itself is something of a misnomer, popularised by Lady Charlotte Guest, the Victorian aristocrat who first translated them into English. But of the eleven stories in the collection, the first four do share a narrative arc: they are known as the 'Four Branches of the Mabinogi' – the very structure 'suggesting the image of a tree', according to its most recent translator, Sioned Davies.[4] The 'Fourth Branch', in particular, is full of references to oak trees. And, unlike some legendary lands, many of the places named in *The Mabinogion* correspond to real-world locations. The rain-soaked Atlantic rainforests of Wales were the backdrop to these myths.

One of the central characters in the Fourth Branch – indeed, one of the central figures in ancient Welsh mythology overall – is the shape-shifting wizard Gwydion. He was a hero of truly cosmic proportions, spanning the world and the Otherworld, and possessed of a deep, dark magic. The Welsh name for the Milky Way galaxy is *Caer Wydion*: Gwydion's Fortress. But the name Gwydion itself is more earthly: it can be translated as 'wood-knowledge', 'wood-wise' or, possibly, 'born of trees'. In fact, the Welsh word for 'trees' – *gwydd* – shares a root with the words for 'knowledge' and 'consciousness'.[5]

In Gwydion, there is an echo of the Celtic druids: 'druid', after all, means 'oak-knowledge' or 'oak-seer' (*derw-weid*). We know that druids existed in north Wales at the time of the Roman conquest, around AD 60, because the Roman historian Tacitus recounts how the invading army encountered them on Ynys Môn (Anglesey). Chanting druids, he writes, were 'pouring forth dreadful imprecations', which 'scared our soldiers'. We also know they considered the Atlantic oakwoods sacred: Tacitus tells of how the druids' 'groves, devoted to inhuman superstitions, were destroyed' by the invaders.[6] Gwydion seems to be a cultural memory of this: he was a wizard of Wales' rainforests.

In the Fourth Branch of the Mabinogi, Gwydion tricks Pryderi, lord of the southern Welsh kingdom of Dyfed, into handing him some magical swine that were gifts from the faerie kingdom of the Otherworld (*Annwn*). Gwydion conjures twelve hunting dogs, and twelve fine stallions with golden saddles, to give to Pryderi in exchange for the swine. But the horses and hounds are in reality made out of toadstools, and, like faerie gold, revert to being toadstools after one day. Before the magic wears off, Gwydion scarpers with the swine, headed back to north Wales. Furious at the deception, Pryderi musters an army and sets off in hot pursuit of the trickster wizard. They clash at Y Felenrhyd: the site of the rainforest Louisa and I visited with Jon and Emma.

The Mabinogion recounts that 'as soon as they reached Y Felenrhyd . . . the foot-soldiers could not be restrained from shooting at each other'. In my mind's eye, I see two armies stalking one another through the misty Atlantic rainforest like guerrilla forces in the Vietnamese jungle, hiding behind moss-encrusted oaks to fire off arrows. Gwydion decides that, rather than risk any more men – this is a war over pigs, after all – he will instead fight King Pryderi in single combat. So they fight; 'and because of strength and valour, and magic and enchantment, Gwydion triumphed and Pryderi was killed'.[7] Wood-knowledge, the magic of the rainforests, could slay kings.

There is some disagreement, however, over where Pryderi's fallen body was buried. *The Mabinogion* says he was laid to rest in the village of Maentwrog, 'above Y Felenrhyd, and his grave is there'; a standing stone in the local churchyard is traditionally assumed to mark his burial spot. But another ancient Welsh poem, 'The Stanzas of the Graves', suggests the slain king was entombed in the oakwoods themselves, where a small stream flows into the Afon Prysor at Ivy Bridge, near the entrance to Coed Felenrhyd. When we passed this spot on our visit, I noticed all the surrounding trees to be swathed with the green scales of tree lungwort. Intriguingly, a nearby farmstead is called Tyddyn Dewin: the 'wizard's smallholding'.[8]

This is far from the only story involving Gwydion and oaks. Later on in the Fourth Branch, the wizard uses his protean powers to fashion a woman out of blossom. Working with another magician, he takes 'the flowers of the oak, and the flowers of the broom, and the flowers of the meadowsweet, and from those they conjured up the fairest and most beautiful maiden that anyone had ever seen. And they baptized her . . . and named her Blodeuedd ["flowers"].' The straw-coloured catkins of the oak, chrome-yellow petals of broom and white froth of meadowsweet can all be found in abundance in Wales' rainforests in spring.

Gwydion intends Blodeuedd to be a wife for his nephew, Prince Lleu Llaw Gyffes. But she falls in love with another, Gronw Pebr, lord of Penllyn, and the lovers conspire to kill Lleu. Gronw strikes Lleu with a poisoned spear, who flees by turning into an eagle and flying away. As Gronw and Blodeuedd seize power in a coup, Gwydion, horrified by the betrayal, sets out to find the wounded prince. He tracks him down to a place called Nantlleu ('Lleu's valley'), in mountainous Snowdonia, deep in the heart of Wales' temperate rainforest zone.[9]

Gwydion chances upon a sow feeding on rotting flesh and maggots beneath an oak tree. Looking up into the tree's branches, he realises that the pig is eating the entrails of Lleu, who perches, eagle-formed, in the oak's canopy. To tempt him down, Gwydion sings a poem, called an *englyn,* followed by a second and a third. Each of the three poems begins by referring to the oak tree; clearly evoking the topography and climate of Welsh rainforests, as well as their supernatural qualities:[10]

> *An oak grows between two lakes,*
> *Very dark is the sky and the valley . . .*

An oak grows on a high plain,
Rain does not wet it, heat no longer melts it . . .

An oak grows on a slope
The refuge of a handsome prince . . .

Lleu's transformation into an eagle and escape into the mountains of Snowdonia says something else about the ecology of Wales at the time the stories of *The Mabinogion* were first sung by its bards. The original Welsh name for Snowdonia is *Eryri*, which many translate as meaning the 'abode of eagles'. Golden eagles have been all but extinct in Wales since the middle of the nineteenth century, shot and persecuted by Victorian gamekeepers. Only recently has an eagle reintroduction plan been hatched.[11]

After tempting Lleu down from his *eyrie*, Gwydion sees that his wounds are tended to and that he recovers his strength. Lleu Llaw Gyffes then takes his revenge by killing Gronw Pebr and reclaiming the throne. Gwydion, meanwhile, transforms Blodeuedd into an owl, the bird hated by all other birds. *Blodeuwedd* in Welsh means 'flower-faced': an apt description of the feathered face of an owl.

I know this story deep in my bones, because it's one I've heard since I was young. It's the plotline to Alan Garner's magnificent children's book *The Owl Service*, in which the three central characters – Gwyn, Alison and Roger – find themselves re-enacting the love triangle of Lleu, Blodeuedd and Gronw in the modern setting of 1960s Wales. After Gwyn finds a stack of old plates in the attic, bearing a floral pattern that also seems to resemble an owl – the 'Owl Service' of the novel's title – the mythical cycle is triggered anew. The trio find themselves increasingly trapped in their destinies, and the claustrophobic landscape itself seems to be conspiring against them.

Reading the book as a kid made me shiver with delight,

and not just for its pacy horror: the setting of north Wales was one I knew from holidays spent happily roaming the hills and paddling in streams. The book was also my first introduction to the politics of Welsh independence, Alison's and Roger's English snobberies towards the Welsh Gwyn only adding to the book's tension. But it was the character of the estate handyman, Huw Halfbacon, whom I found most intriguing. He is revealed to be Gwydion the wizard, who mumbles to himself regretfully about unleashing Blodeuwedd: 'She wants to be flowers and you make her owls . . . Mind how you're looking at her.'[12]

One last ancient tale exists which speaks to the magic of the Welsh rainforests. The *Cad Goddeu*, or 'Battle of the Trees', is a surreal epic poem in *The Book of Taliesin* – another collection of Welsh stories, which may date back as far as the sixth century AD, not long after the Romans left Britain.[13] In the *Cad Goddeu*, the world is threatened by an invasion of hideous beasts from the Otherworld, including a 'many-scaled monster / With its hundred heads' and a 'black forked toad / Of a hundred claws'.[14]

So Gwydion, described as the 'great magician of Britain', enchants a forest and raises up an army of trees to march into battle. Centuries before Shakespeare had Birnam Wood come to Dunsinane in *Macbeth* – and long before Tolkien wrote about the march of the Ents upon Isengard in *The Lord of the Rings* – here is a poem about 'when trees went to war'. Parts of the poem certainly have the quality of a fantasy movie:

> *When the trees were enchanted . . .*
> *They mowed soldiers down*
> *With their mighty boughs.*

Most of the poem, however, is mysterious to the point of being baffling. The central section describes a succession of warrior-trees and soldier-plants, each imbued with human characteristics.

> *First came the Alder,*
> *Which struck the first blow . . .*
> *Oak's passionate shout*
> *Made earth and sky shake.*

Some commentators have refused to read this part as being mere fantasy, however, and asserted that it must contain a hidden meaning. The writer Robert Graves, in his influential work on pagan mythology *The White Goddess*, speculated that each of the trees represented the letters of the original ogham alphabet of the Celts. But as the poem's most recent translators, Gwyneth Lewis and Rowan Williams, have said, this is 'now generally agreed to be fanciful'.[15]

I wonder if there's a simpler explanation: that the 'Battle of the Trees' is a naturalist's almanac, containing a guide to the flora of western Britain. At least thirty-three species of trees and plants are mentioned, often with realistic attributes. The account of the 'spiny Blackthorn' as being 'hungry for bloodshed', for example, is an accurate description of what happens if you ever tangle with a sloe bush; likewise the lines that the 'Infamous Hawthorn / Gave festering wounds'.

The order in which the trees are introduced might even reflect the time of year when they first come into flower. Alder, for instance, which 'struck the first blow', is an early source of nectar for bees, as its catkin flowers can appear as soon as February. Meanwhile, 'Birch, although noble, / Was slow to get dressed', perhaps alludes to its dressing of catkins only appearing in April or May. Rowan, who arrives 'late to the muster', doesn't come into flower until May–June.

And whilst trees and shrubs don't ordinarily get up and walk, they're certainly known to 'invade' open ground through the process of natural colonisation. One passage in the poem brackets together all the main species that today we term 'scrub' in English, and, in Welsh, *ffridd*:

> *Bracken grew rampant;*
> *But broom, at the head,*
> *Was trampled in mud.*
> *Though unlucky, Gorse*
> *Still joined in the force.*
> *A spell brought Heather*
> *To join, famous fighter.*

To me, this sounds like a botanists' roll call of the plants that make up the scrubby, heathy vanguard of a forest as it advances up a hill.[16]

My cod theories aside, I needed some help understanding the place of Welsh rainforests in this myth and magic. So it was wonderful serendipity when, whilst I was in Wales, an email pinged into my inbox from someone who could do just that. It was none other than the translator of *The Book of Taliesin*, the poet and writer Gwyneth Lewis – the first National Poet of Wales. She was captivated by my lost rainforests project and wanted to assist.

The words of Gwyneth's poetry are emblazoned on the Welsh Millennium Centre in Cardiff, immortalised in glass and steel: *In These Stones Horizons Sing*; and *Creu Gwir fel Gwydr o Ffwrnais Awen*, which translates as *Creating Truth Like Glass from Inspiration's Furnace*. Gwyneth has said that her choice of these lyrics refers in part to *The Book of Taliesin*, in which the bard Taliesin receives inspiration from an enchantress's copper cauldron.[17]

Gwyneth was inspired by my rainforests project partly because there is a fragment of rainforest near where she lives and writes in the Ffynnon Dewi valley on the coast of Ceredigion. She told me that walking down into this tree-lined gorge is like entering 'a green velvet Gothic cathedral'. The good portents didn't end there. 'I'm writing from the converted pigsty of a farm called Pendderw'– a word which translates as 'head of the oak' – 'so I'm right in the middle of the Fourth Branch of the Mabinogi, with Gwydion following a sow to find his wounded nephew, Lleu, eagle-formed, hiding at the head of an oak tree.'

When we got to speak, I wanted to ask Gwyneth whether I was in danger of romanticising Welsh myths and their meaning to modern Wales.

'Oh, don't fall for any Celtic bullshit!' she warned me. We discussed how a lot of hippy-dippy nonsense is written about druids and Celtic religion, much of which is simply unknown. There is also, I'm acutely aware, a long history of English condescension towards Welsh culture, which likes nothing better than to patronise it as mystical but marginal: the 'Celtic fringe' – or, as Jon Moses puts it, a form of 'Welsh orientalism'.

But Gwyneth also reassured me that I was on to something in looking for signs of lost rainforests in Welsh legends. 'I'm not a romantic,' she said to me, 'but I'm very aware of the metaphorical depths in certain Welsh place names. I'm wary of making special claims about Welsh people having a particular knowledge of nature; we're not the original Brythonic Celts, we're all immigrants, too. Yet the links between mythology and the landscape are much closer to the surface in Wales.'

That, to me, is the surest message we can take from these wonderful myths. As a tree-hugging environmentalist, I'm in sympathy with tales of Celtic tree-worship, of druids and wizards revering sacred rainforest groves. But what draws me to them isn't the hope of discovering some hidden code or fragment of pagan wisdom. It's that these stories contain a deep love of place, infusing the real world with sacred meaning. In a time of ecological crisis, that's a story we badly need to relearn.

*   *   *

The remaining four days of our Welsh rainforests road trip took us through countless magical places.

We drove south to Coed Crafnant, on the edge of the Rhinog mountains, where we wild-camped under the stars, staring up at Gwydion's Fortress and the ripening moon. Jon slept inside his bivvy bag, in a hammock slung between two trees, whilst Louisa and I pegged out our tent on the mossy banks of the Afon Artro. The next morning, the three of us took a bracing dip in the ice-cold waters of the river and made coffee in a pot perched on our gas stove, before exploring the rainforest on our doorstep.

The visitor noticeboard put up by the Wildlife Trusts explained that Coed Crafnant was a 'wild, ancient Atlantic oak woodland . . . steeply sloped and roughly divided into terraces, each with its own microclimate'. Tree lungwort bloomed in awesome profusion on oaks and ash trees. Stone walls leaned drunkenly, apparently collapsing under the accumulated weight of moss. A tiny frog, no bigger than a twopence piece, leapt off the footpath to escape our boots. A sand-coloured lizard basked on a rock, its tail missing, lost to an unknown predator. Birds skitted between the trees: the brown back of a treecreeper, the scarlet breast of a redstart.

The rainforest at Coed Crafnant borders the Rhinogs, an area of upland the likes of which I've seldom seen elsewhere in Britain. The Gwynedd Archaeological Trust calls it 'an example of upland vegetation only lightly influenced by grazing'.[18] We saw no sheep, only occasional packs of feral goats – unmistakeable by their smell, like sour milk. As a result, the moorland was bursting with life: thousands of flying ants, bees and hoverflies feeding on the purple heather and orange bog asphodel, red admiral butterflies flickering from plant to plant. With few livestock to nibble them, rowan saplings were colonising the edges of the bogs. Bilberries fruited in abundance, and we ate them as we climbed, their dark purple juice staining our fingers like woad.

We drove on, continuing south. At Coed Simdde-lwyd, another Wildlife Trusts reserve in the Vale of Rheidol, Jon found the orange fruiting bodies of a fungus which looked suspiciously psychedelic, and furtively stashed some in an empty Pringles tube. The more mushrooms Jon foraged, the more I came to think of him as our trip's equivalent of Gwydion the magician: tapping into some psychoactive, mind-bending mycelial network just as the wizard used his wood-knowledge to transform toadstools into stallions.

'There's probably something you can tell about the health of a rainforest by the number and diversity of fungi in it,' said Jon, as he peered at the umpteenth mushroom. 'Something to do with the age and size of the mycorrhizal networks in the forest soils, and whether they've been disturbed much.' I'm sure this is true. Mycologists have found that mycelial networks interwoven with tree roots can enable trees to pass information between one another, such as warnings about predators, forming what's been termed the 'Wood Wide Web'.[19] In some of Wales' last rainforests, their woodland floors potentially undisturbed for thousands of years, the networks are doubtless rich and deep.

Our winding road trip took us to unexpected places that confirmed to me the profound role rainforests have played in Welsh history. In a dark corner of Y Sospan café, in the small town of Dolgellau, we found an astonishing clue to the temperate rainforests that once carpeted Wales. On the café's first floor, hiding near ceiling-height amongst old beams riddled with woodworm, was a seventeenth-century plaster frieze depicting an ancient hollow oak tree which once stood in the nearby Nannau Estate. The story goes that Owain Glyndwr – the dashing Welsh prince who rebelled against the conquering English in the early 1400s, a Che Guevara of his day – once slew a rival and hid his body in this hollow oak, and that the corpse's spirit haunted the spot forever after.

The frieze was an extraordinary piece of work, and not just

for the grotesque figures that populated it: the Devil dancing in the branches of the oak; a brigand being hanged from the ivy-clad branches; a witch on trial beneath it being drowned in a ducking pond. Their faces leer cartoonishly, like Punch and Judy puppets on a seaside pier. But they weren't what really drew my eye. What most interested me was how the artist had depicted the old oak: it was covered in lichens, carpeting it in three-dimensional fur.

As far as I'm aware, this is the earliest depiction of epiphytic plants in any work of art in Britain. It shows not just a legendary old oak, but a rainforest tree. It suggests that Glyndwr, who waged a guerrilla campaign against Wales' English imperial overlords, did so from the fastnesses of the Welsh jungle.

Yet when we drove up the road from Dolgellau to visit the Nannau Estate where Owain's oak once stood, it became clear that something was wrong. The hollow tree was long gone, zapped by a lightning strike centuries ago. But something else had since been consuming the rainforest. As we walked through the estate's wet woodlands, admiring the lichen-clad oaks dripping with polypodies, we came face to face with the single greatest threat to Wales' rainforests; something that has come to symbolise the modern Welsh landscape, even Welsh culture.

A sheep.

It bleated, turned, and scampered off in search of fresh saplings to eat. All around us, veteran oaks and rowans creaked and groaned, slowly succumbing to senility. There was no sign of a new generation of trees coming forth to replace them. They had all been eaten. In their place, invasive rhododendron bushes were now moving in to finish the destruction that the sheep had started.

This creeping devastation isn't just an ecological catastrophe for Wales' rainforests. It's also, I'd argue, an act of cultural vandalism. Wales' history and myths, as we've seen, are replete with references to the Atlantic rainforests that once carpeted the ancient Welsh kingdoms, and still cling on in places. But now this heritage is being nibbled away, like a beautiful tapestry being eaten by moths.

What's even worse is how few realise what's happening. To most people, Wales is not a land of rainforests, but rather a land of green grass, rolling hills, stone walls, and shepherds. A land, in other words, that's been made by sheep. As we beat a hasty retreat from the sheep-riddled woods of the Nannau Estate, a sign by the roadside confirmed this in no uncertain terms. It warned, ominously: THIS IS SHEEP COUNTRY.

To an ecologist, the wet climate of Britain's west coast is rainforest territory. But seen through the eyes of many farmers, this lush and verdant land is good for one thing above all else: pasture.

It's a difference captured in the UK government's National Food Strategy, published in 2021, which noted a common refrain amongst farmers: 'We have an excellent climate for growing grass.' But, the strategy went on to argue, 'although ruminants are experts at turning grass into delicious meat, that is not the only useful purpose this land could serve . . . Many upland areas were once covered in temperate rainforests. Now they are covered in sheep, which nibble very close to the ground and make it difficult for tree saplings to grow.'[20]

To date, such arguments have fallen on stony ground in Wales. This is a country where sheep outnumber people three to one: the latest statistics from the Welsh government put the number of sheep in Wales at just shy of 9 million, whilst the human population stands at only 3 million.[21] Only New Zealand, Mongolia and the Falkland Islands have a greater ratio of sheep to people.[22]

Yet the dominance of sheep is comparatively recent in Wales. You'll search in vain for any mentions of sheep in the Four Branches of the Mabinogi (although they do crop up in two of the other tales recorded in *The Mabinogion*). Pigs, on the other hand, appear on numerous occasions – from the magical swine stolen by Gwydion to the sow that ate Prince Lleu's rotting flesh. As translator Sioned Davies notes, pigs 'are often associated with the supernatural in the *Mabinogion*, and played a significant role in the mythology of the Celts'.[23] Pigs rootle around in the earth and cause disturbance, but they do far less damage to regenerating woods than sheep; indeed, wild boar are creatures of woodlands. Early Wales, it seems, was a land of swine and rainforests, rather than sheep and grass.

Sheep, in fact, were a foreign import to Wales. First domesticated in Mesopotamia, sheep were probably present in Britain as early as the Neolithic, but weren't farmed at scale until the Middle Ages. Large-scale sheep ranching only began when the Cistercian monasteries started colonising the remote uplands of central Wales in the 1100s. The Cistercians were a monastic order founded in France, whose spread to Wales transformed the physical landscape as much as the spiritual life of the country. Abbeys like Strata Florida, built on the edge of the Cambrian mountains, quickly established vast granges in the heart of Wales' rainforest zone and began farming the land for the increasingly lucrative wool trade. What rainforests still existed in the Cambrians at this stage would surely have succumbed to the monks tending their flocks.[24]

It would be impossible to deny, however, the centrality of sheep to modern Welsh farming. The harsh climate of Wales' windswept, treeless hills, coupled with the hardiness of many modern sheep breeds, means they have become a mainstay of Welsh agriculture, accounting nowadays for a fifth of the country's agricultural output.[25] And though the bucolic image of the small-scale farmer is dying out in much of Britain as mega-farms proliferate, in Wales, many family run hill farms persist. Thanks for this is due to another Welsh Wizard: not Gwydion this time, but David Lloyd George, land reformer and British Prime Minister around the time of the First World War. His taxes on wealthy landowners led to the break-up of many large Welsh estates, with much land sold to the former tenants, transforming Wales into a country of small-scale owner-occupier farms. The sturdy sheep was now joined by the self-sufficient yeoman farmer to create a powerful symbol of proud Wales.

It fell to a modern Taliesin to turn this new economic reality into a national mythology. R. S. Thomas, perhaps the greatest poet of twentieth-century Wales and an ardent Welsh nationalist, deified the Welsh shepherd. For Thomas, as for many people in Wales today, the hardy shepherd was the nation's folk-hero, a fundamental mainstay of Welsh culture.

In his 1946 poem 'A Peasant', Thomas describes 'Just an ordinary man of the bald Welsh hills, / Who pens a few sheep in a gap of cloud'. Though Thomas sneers at the shepherd being 'half-witted', with 'something frightening in the vacancy of his mind', he also celebrates him as 'a winner of wars', who 'preserves his stock, an impregnable fortress', telling his Welsh readers: 'this is your prototype'.[26] There's no need to swallow Thomas's condescending, *völkisch* portrayal of Welsh shepherds as noble savages. But there's little doubt that Welsh farming communities helped preserve the Welsh language when it was on the verge of dying out, and who resisted – as Thomas put it in another poem – 'the English / Scavenging among the remains / Of our culture'.[27]

Reading R. S. Thomas raging about what the English have done to Wales, it's hard not to weep with shame. 'There are places in Wales I don't go: / Reservoirs that are the subconscious / Of a people,' he wrote in the aftermath of the Tryweryn debacle, when a village was flooded to supply Liverpool with drinking water.[28] Thomas was rightly furious at the drowned valleys and the farmsteads engulfed by forestry plantations to supply England with water and wood. This was resource extractivism pure and simple, with Wales treated as 'England's first and final colony', as Adam Price, the current leader of Plaid Cymru, has put it.[29]

Out of this context has arisen the antipathy of many Welsh farmers to rewilding. 'Rewilding' is the large-scale restoration of ecosystems to the point at which nature can take care of itself, reinstating natural processes and reintroducing missing species: think of the now-famous Knepp Estate in Sussex, or the various efforts to bring back the beaver.[30] I first came across the concept when living in the town of Machynlleth in mid-Wales, during which time I became friends with a resident whose work introduced me to the existence of temperate rainforests in Britain: the writer George Monbiot.

For decades, 'Mach' has been something of a Mecca for environmentalists, many of them English, drawn to the town by the Centre for Alternative Technology, which was founded in a nearby quarry by hippies in the 1970s. When I lived there in the early 2010s, George was writing *Feral*, his clarion call to rewild Britain's denuded landscapes. It drew on his experiences of walking in the hills around Mach, sinking into depression as he realised most were bereft and empty of life. 'In mid-Wales, I found, the woods were scarce and, in most cases, dying, as they possessed no understorey,' George wrote. The cause, he concluded, was that they had been 'sheepwrecked'.[31]

George's words certainly resonated with me, and on my own hikes around the local area I began to see signs of decay and senescence in the landscape. The lone trees clinging on to the windswept hills were now reminders of a long-gone ecosystem; the bright green pastures no longer looked so pleasant. The early environmentalist Aldo Leopold described perceiving the world through an ecologist's eyes as seeing 'a world of wounds'. It's certainly one way to ruin a good walk.

But clearly not everyone sees in this way; particularly not some Welsh farmers, who've worked the land their whole lives, and come to love it as it is. In *Feral*, George met with the sheep farmer Dafydd Morris-Jones, who farms in the heart of the Cambrian mountains, the 'Green Desert' that dominates central

Wales. 'My concern with rewilding is that it takes the people out,' Dafydd told him. 'If you undermine the core economies that support the Welsh-speaking population in the language's heartland, you write us out of the story.'[32] To farmers like Dafydd, rewilding is the latest form of colonialism foisted on Wales by middle-class English incomers. It may not seek to suck resources out of the country like forestry and reservoirs do, but by seemingly threatening the livelihoods of the small-scale sheep farmers who have kept the Welsh language alive, it's seen as an attack on Welsh culture.

Some time later, after I had left Machynlleth, the town and surrounding hills found themselves at the centre of this 'culture war' pitting sheep farmers against rewilders. A rewilding project called Summit to Sea had been proposed for the area by a coalition of environmental NGOs. Some of these groups were deeply rooted in the local community, and some of them were seen as 'outsiders', such as the organisation Rewilding Britain, set up by George's partner Rebecca Wrigley following the publication of *Feral*.

Summit to Sea quickly met a wave of vitriolic opposition from Welsh farmers and farming unions. 'It shouldn't exist – it shouldn't be here,' thundered the Farmers' Union of Wales.[33] 'There was no discussion, no consultation, no prior notice,' hill farmer Iwan Pughe Jones told the *Farmers Guardian*. 'This is not a collaboration; it feels like an attempt to override our rural community.'[34]

Years after the Summit to Sea debacle, I worked for Rewilding Britain on campaigns, and commissioned some polling into public attitudes to rewilding. Given how much the media and farming unions portray rewilding as a polarising idea, I was surprised to find that an overwhelming 81 per cent of Britons supported it – with just 5 per cent opposed, and 14 per cent undecided.[35]

Of course, such support should never be taken as justification for riding roughshod over the concerns of those opposed. Any proposal to fundamentally change the character of a local area – no matter how beneficial such changes might be for both people and place – requires careful consensus-building.

And in the hills around Machynlleth, feelings about rewilding still run strong, as I found when we visited on our rainforests road trip. NO TO REWILDING, bellowed one sign in a farmer's field on the outskirts of the town. The war of words, rewilding versus sheep, grass versus rainforests, rumbles on.

But we hadn't come here looking for conflict. We'd come in search of hope. Joe Hope, in fact.

Joe Hope is a rare breed: an environmental scientist who's become a farmer. He has a doctorate in forest ecology, but now farms pigs and cattle in the hills south of Machynlleth. He is, at present, something of an anomaly. But the farming he practises represents both a return to old traditions and a window on to a possible future.

We sat in Joe's farmhouse kitchen, drinking tea out of mugs with pictures of cows on them. Corn dollies hung from the dark ceiling timbers next to sticky fly-paper. Two beautiful tabby cats prowled around, looking for free strokes. Unusually for a farmhouse, there were no dogs: Joe is 'definitely a cat person' – I told him I approved – and with no sheep to herd has no need for them on the farm. Plates and crockery were neatly stacked in a drying rack above the sink; this was a shared house where friends and neighbours were always popping in. Joe talked about ordinarily being 'hyper-social' and struggling with Covid lockdowns, though he added that he was glad to be based here, amidst nature, for the duration of the pandemic.

Joe Hope describes himself as an optimist. He's trying to bridge the increasingly polarised divide between farmers and environmentalists. Though born in England, he's lived most of his life in Scotland, but now says he feels most at home in

Wales. To some, no doubt, this still makes him an 'incomer'; but as Gwyneth Lewis, the first National Poet of Wales, reminded me, 'we're all immigrants'.

What distinguishes Joe from Machynlleth's population of more transient hippies – who, like me, seldom learned much Welsh during their stay in the area – are his efforts to integrate. He speaks Welsh, and has been researching the old field names on his farm, seeing what he can discover about their traditional uses. He keeps on good terms with other farmers in the area, sometimes investing in shared farm machinery; as Joe showed us around his small forty-acre farm, a neighbour passing by in a Land Rover gave a friendly wave. 'I've got huge respect for the hard graft that all farmers do around here,' he told us. As a result – despite his unusual farming methods – Joe is increasingly accepted.

As Jon, Louisa and I leaned on a fence, watching the chickens peck around the yard, Joe wandered off carrying a bucket of feed. A few minutes later, a mournful, ululating cry pierced the air.

'*Cooos . . . Cooooooooooooos . . .*'

It was Joe, calling to his Highland cattle up on the hillside above the farm. 'I'm trying to train them to come when called,' he explained. 'Helps for herding them and bringing winter feed.' The herd, their long horns held proudly aloft and their hair as ginger as Joe's, slowly trotted down the hill towards their master. Highland cattle, whilst obviously not native to Wales, are closer to the ancient aurochs that once inhabited Britain than most modern breeds. They graze the grass, but not as close to the ground as sheep, allowing more species of plants and wildflowers to return.

In a field opposite, we were introduced to Joe's pigs, who came gruffling and snouting through the bracken and rowan trees that covered the steep ground. Pigs are well suited to wood pasture: they trample the bracken, letting in light, and the way they churn up the soil can be excellent ground preparation for seedlings to take root. Cattle and swine, but no sheep: a return to an older way of Welsh farming, of the sort celebrated in *The Mabinogion*, before the woolly incomers arrived. Like R. S. Thomas's idealised peasant, Joe, too, is a 'man of the bald Welsh hills'. It's just that his way of farming is making the hills a little less bald.

In other ways, as well, Joe is trying to break conventions. He wants to start up an oat-milk business, growing and milling oats sown using heritage seeds that are genetically more diverse than the varieties sold to farmers by seed-dealers. By selling both oat-milk and beef, Joe wants to cater for both vegans and meat-eaters – bridging another divide that has crept into our polarised public discourse around food. Some farmers rage against veganism, and refuse to reduce livestock numbers despite the damage done by overgrazing; but Joe reasons that vegans eat, too.

Having now given us a tour of the farm, Joe was excited to show us something else nearby. 'I know a brilliant place for lichens just down the valley,' he told us with glee, his eyes lighting up. 'A real Lobarion hotspot.' We bundled into Zoomer and drove down the potholed forestry tracks that are Joe's only route out to civilisation. 'Let's pull up over there,' he gestured after a while, so we parked up, unbuckled our seatbelts and walked up the dirt track into the forest. Some of the forestry plantations around here, I knew from past explorations, contained the hollow shells of old farmsteads, dynamited by the Forestry Commission to make way for its armies of Sitka spruce.

The overgrown ruins that Joe took us to, however, were industrial in nature. Amidst the monoculture of pines was a tumbledown collection of walls, and within them sprang ash trees, hawthorns, rowans, oaks. And on these trees, lichens bloomed in rambunctious profusion.

For some reason, Joe had brought a houseplant spray with him. 'Why . . .?' I asked. 'Because lichens look at their best when they're wet,' he said, giving a patch of tree lungwort a casual squirt, so that it glowed green. My geekish excitement at the rare species we were seeing was as nothing compared to Joe's expert enthusiasm. Lichen names tumbled from his mouth like the cataracts of water cascading down the valley stream below. 'This one's *Peltigera*, dog lichen . . . this dark stain is some *Parmeliella* . . . and look, this is very unusual – it's lungwort, but growing on the very tips of the branches, rather than just on the trunk of the tree. It must really love it here.' It was wonderful to see both sides to Joe, the farmer and the ecologist; different aspects to his character, bringing different perspectives, but equally enchanted by the world.

Why were there so many rare rainforest lichens in this one tiny patch, I wondered. Joe had a theory. 'This used to be a lime kiln,' he said, pointing to the ruins. 'The stream down there is still called Nant Y Factory, the factory brook. Lichens, particularly rainforest lichens, thrive in alkaline conditions. It's why so many lichens died out during the Industrial Revolution – all that acidic, sulphurous smoke from factories killed them. But this was a factory producing lime, which is alkaline. It's possible that the lime left in the soils has made the tree bark alkaline, creating a hotspot for rare lichens.'

I stared at the factory ruins and their effervescence of lichens with growing excitement. We were standing in a post-industrial rainforest. Where once this valley had been scarred by industry a rainforest had now returned. And though ours was only a tiny patch, nestled in a swathe of plantation, it seemed to hold out the promise of redemption. I felt uplifted. If a rainforest can thrive on the wastes of industry, then surely they can be coaxed back on to some of Wales' sheepwrecked hills, too.

*   *   *

My sense of optimism later faded when we drove the last leg of our road trip, homeward-bound, through the Cambrian mountains. The road wound steeply up past oakwoods and conifer forests, stone walls and rocky outcrops. Then we rounded a corner into a valley, and my heart sank.

Before us stretched mile after mile of grey-green sheep pasture, relentless in its monotony. The sky lowered overhead, darkening to a gunmetal grey in sympathy with the scarred land. It was the bleakest landscape I have ever seen; apocalyptically bleak. Louisa, Jon and I sat in stunned silence as we drove through it, occasionally muttering disbelieving swearwords. After a week of euphoric adventures in Wales' rainforests, this was quite the comedown.

This was the very heart of Wales' rainforest zone, where the oceanic climate conspires to make conditions perfect for the rich profusion of plant life that we'd spent the past week exploring. Yet here, humanity had found a rainforest and turned it into a desert. It had started long ago, no doubt: Wales' Green Desert is the product of agricultural malpractice dating back to the twelfth-century monks of Strata Florida. But what began as a profitable enterprise in medieval times today supports a mere twenty-eight farms over an area covering 46,000 acres.[36] The farming unions claim rewilding will lead to rural depopulation, but centuries of overgrazing have already drained this land of both people and wildlife.

And, in doing so, Wales is losing a part of its heritage, its culture, itself. Because the Wales of this great country's myths and legends was a rainforest nation, whose peoples lived amongst and coexisted with the Atlantic oakwoods that once carpeted their land, celebrating them in song. They knew these rainforests and knew them deeply, weaving them into their stories, vesting their greatest heroes with a magic derived from that profound knowledge of place and ecology.

There is a way back to this, but it is unlikely to come through a culture war between sheep farmers and rewilders. The truth is that there is more than enough space in Wales, as there is in the rest of Britain, both for farming to continue and for more rainforests to flourish. But it needs to be a different type of farming: fewer sheep, a shift to cattle and swine, and more space for nature to thrive on the least productive land. Farmers like Joe Hope are showing one way to this future. Poets like Gwyneth Lewis are showing another. Because for Wales to throw off its self-image as a land of sheep and grass, and rediscover its history as a land of Atlantic rainforests, it first has to feel inspired to do so. It needs to be re-enchanted with the magic of the rainforests.

6

A Perfect Republic of Shepherds

To the Romantic poet William Wordsworth, the English Lake District was a 'perfect Republic of Shepherds'.[1] We think of him today as a poet of daffodils and sheep; as a whimsical wanderer who celebrated the farmed landscapes of the Lakes as a perfect union of humanity and nature. Yet there is another side to the great poet, one which posterity has buried under a mountain of tourism brochures and heritage guides. William Wordsworth was also a tree-hugging environmentalist, who bitterly lamented the deforestation that humankind had wrought upon the Cumbrian fells. You might call him our first poet of the rainforests.

'Twice as much rain falls here as in many parts of the island,' Wordsworth noted in his *Guide to the Lakes*, which ran through several editions between 1810 and 1835.[2] To Romantic poets like Wordsworth, Coleridge and Keats, the oceanic climate of the Lake District only added to its drama and beauty.[3] It certainly added to its native flora, which Wordsworth described with awe and wonder. He celebrated not only the 'brilliant and various colours of the fern', but also 'the lichens and mosses: their profusion, beauty and variety exceed those of any other country I have seen'.[4] In Wordsworth's *Guide*, we also find one of the earliest written descriptions of epiphytic plants. He related how the 'minute beauties' of lichens and mosses were 'scattered over the walls, banks of earth, rocks, and stones, and upon the trunks of trees, with the intermixture of several species of small fern, now green and fresh'.[5]

Wordsworth was acutely aware that the Lake District had been deforested. 'Formerly the whole country must have been covered with wood to a great height up the mountains,' he wrote, recounting that: 'When the first settlers entered this region . . . they found it overspread with wood; forest trees, the fir, the oak, the ash, and the birch had skirted the fells, tufted the hills, and shaded the valleys, through centuries of silent solitude'.[6]

He was quite right to claim this. Like Dartmoor, the Lake District is now haunted by many ghost woods, their fallen boughs memorialised only in old place names. The River Derwent – which runs through the damp heartland of the Lakes – derives its name, like the Dart in Devon, from the Brythonic Celtic word for oak, *derw*. Later, during the Dark Ages, the Lakes were settled by Viking invaders, who added their own descriptions of the landscape. The Celtic worship of the oak was replaced by the Vikings' reverence for ash trees: Norse mythology held that a titanic ash, Yggdrasil, supported all the realms of the cosmos in its roots and branches. *Askew*, *Askrigg* and *Ashgill* are places in the Lakes named after ash trees, and many of the temperate rainforests in Cumbria have a preponderance of ash.[7] Birker Fell, to the southwest of Scafell Pike, gets its name from the Old Norse for 'birch hill'. Today, it is entirely bereft of trees, birch or otherwise.[8] Ennerdale, to the west, comes from the Norse for 'juniper valley'. The juniper trees, with their bittersweet berries used for flavouring gin, have only recently started to return after centuries of absence – under the auspices of the National Trust, who nowadays own about a fifth of the Lake District.[9]

Oak, ash, birch, juniper: all were cleared by successive waves of Lakeland settlers to make way for agriculture and industry. Wordsworth knew this. He angrily described the deforestation he had witnessed at a farmstead on the shores of Ullswater: 'The axe has here indiscriminately levelled a rich wood of birches and oaks, that divided this favoured spot into a hundred pictures. It has yet its land-locked bays, and rocky promontories; but those beautiful woods are gone, which *perfected* its seclusion'.[10]

Yet today Wordsworth has been co-opted by heritage fanatics, wheeled out to support the notion that the Lake District is a so-called 'cultural landscape' in which humanity supposedly lives in harmony with nature. In 2017, a group of conservationists and Cumbrian sheep farmers successfully bid for the Lake District to be designated a World Heritage Site by UNESCO. Some environmentalists warned at the time that this move threatened to preserve a degraded ecosystem in aspic: the writer George Monbiot thundered that it would turn the Lakes into a 'Beatrix Potter-themed sheep museum'.[11] The authors of the bid sought to allay such fears by invoking the spirit of the Romantic poets, claiming that 'Wordsworth . . . knew that these places and landscapes were hand-made and managed by a community of hill farmers and shepherds'.[12] But Wordsworth also knew that the hill farmers and shepherds had transformed the natural landscape of the Lakes from one dominated by wet woods to one dominated by grazing sheep.

Rather than an unchanging landscape, the one that Wordsworth described was dynamic and shifting. Today, the famous dry-stone walls that crisscross the lower fells of the Lake District very visibly divide up the landscape for sheep. But as Wordsworth noted, 'when first erected, these stone fences must have little disfigured the face of the country; as part of the lines would everywhere be hidden by the quantity of native wood then remaining'.[13] The Romantic poets could see that the Lake District of their era had undergone a process of ecological decline, even though much beauty in it still remained.

Wordsworth described how 'the beautiful traces . . . of the universal sylvan appearance the country formerly had, yet survive in the native coppice-woods that have been protected by enclosures'; and that 'the forest-trees and hollies, which, though disappearing fast, are yet scattered both over the enclosed and unenclosed parts of the mountains'.[14] Nowadays, this process of tree-loss has gathered pace, with many of the scattered trees and wood pastures described by Wordsworth succumbing to increasingly high densities of livestock. The number of sheep and lambs in the Lakes today is double what it was in the 1950s, let alone what it would have been in Wordsworth's day.[15]

As for the natural fauna of the Lake District, Wordsworth knew that the tamed pastoral idyll was only a recent invention. The original inhabitants of the fells, he wrote, included such wild species as 'the wolf, the boar, the wild bull, the red deer, and the leigh, a gigantic species of deer which has been long extinct; while the inaccessible crags were occupied by the falcon, the raven, and the eagle'.[16] Persecuted for centuries by shepherds, wolves today only appear in Cumbrian place names, from Wolf Crags to Whelpside.[17]

Perhaps most remarkably of all, Wordsworth's writings display an understanding of how the temperate rainforests of the Lake District would regenerate themselves, if only people gave them half a chance. He recounted how a shortage of wood in the seventeenth century had 'made it in the interest of the people to convert the steeper and more stony of the enclosures, sprinkled over with remains of the native forest, into close woods', which, 'when cattle and sheep were excluded, rapidly sowed and thickened themselves'.[18] In other words, Wordsworth was describing how trees and woods would naturally self-seed and spread when not overgrazed by livestock. But in the modern-day 'Republic of Shepherds', this is an argument that often falls on stony ground.

This chapter is about how our rainforests are slowly being nibbled to death by sheep, and what we can do to stop this happening. It's about how some farmers, landowners and conservationists are already taking action to prevent livestock from overgrazing our rainforest fragments – whether by deploying fencing, prickly scrub or the latest satellite tracking technology. But it's also about how our rainforests need to be dynamic habitats, playing host to vibrant natural processes. Seeking to keep any landscape in stasis ends up creating a fundamentally unbalanced ecosystem: what a healthy habitat needs is the right kind of disturbance. An overgrazed woodland is unnatural, but so is a wood without any animals in it at all.

In search of answers to these thorny issues, I decided to make my own visit to Wordsworth country. The daffodils were a distant memory by the time I travelled there in early autumn, but the vales and hills were awash instead with the brilliant red berries of rowan. Arriving in Keswick in the north Lakes, I grabbed a coffee and then boarded a bus heading south into Borrowdale, the damp heart of the Lake District's rainforest zone. As it careened and juddered down narrow roads thronged with day-trippers, I gazed with pleasure at the oakwoods that still clothe the shores of Derwentwater and the river that feeds it. I was headed for one of them in particular: a place called Johnny Wood.

Johnny Wood is a Site of Special Scientific Interest, or SSSI, and a conservationist friend had told me he thought it was 'the most important block of temperate rainforest in England'. As for where it got its unusual name, no one really knows, but it's been called Johnny Wood since at least the seventeenth century:

perhaps Johnny was once its owner or worked it for coppice.[19] The bus deposited me at a crossroads, and I followed the western track past a campsite hemmed by dry-stone walls towards the lush green of the forest ahead of me. Behind it rose the rugged grey mass of Dale Head, breached by the Honister Pass, which holds the UK's rainfall record: an astonishing thirty-four centimetres of rain fell here in a single day in December 2015.[20] People tend to visit the Lake District either to climb its rainswept fells or to admire the rain-fed meres and waters of the valley floors. What's usually missed are the rain-soaked woods that grow between the two.

The sessile oaks of Johnny Wood were taller and straighter than the hunched trees I was more used to seeing in the Westcountry, but their trunks still heaved with moss. Clusters of olive-green mosses jostled for supremacy, pebbledashing the bole of a tree like the skin of a dinosaur. On one oak, I found the biggest beefsteak fungus I'd ever seen: wider than a handspan, the colour of oxblood, and which trembled like jelly when I touched it.

Occasional sounds interrupted my wanderings. A woodpecker knocked hesitantly against a nearby alder, searching for grubs, before letting out a yaffling cry. At one point, a sudden scurrying in the bracken undergrowth betrayed the presence of a mouse or vole. From overhead came the grunt and click of two jousting ravens. But for most of my visit, the wood remained eerily quiet, deathly still.

The scent of death manifested itself in other ways: lots of the oaks in Johnny Wood were dead, or dying on their feet. Dead wood is, of course, an essential part of a healthy, dynamic woodland. It provides great habitat for beetles and their grubs, and hence for the birds that feast on them, like the woodpecker I'd heard. When a dead tree falls, it also creates a clearing in the canopy, letting in the sunlight that understorey plants and many lichens need to thrive. But more worrying was the relative absence of young saplings in Johnny Wood coming through to replace the dead trees.

I soon found out why.

As I looked up, there were two sheep on the footpath ahead, staring back at me. In the schlock-horror movie *Black Sheep*, a hill station in New Zealand is overrun by a plague of zombie sheep, who gaze intently at their human victims before devouring them. Fortunately for me, if not for Johnny Wood, the creatures I'd encountered were definitely herbivores. Perhaps startled by the eye contact, they quickly ambled off into the forest to munch on some saplings. You've heard of rainforest beef, and how it's killing the Amazon: this was rainforest mutton.

Earlier in the year, an ecologist from Natural England had visited Johnny Wood and assessed its condition as being 'unfavourable – declining'. This, to put it mildly, is not a good sign. Out of England's 4,000 SSSIs, just 5 per cent are currently classed as falling into this category.[21] The assessment for Johnny Wood recorded that 'there is a notably low presence of saplings', and found 'widespread evidence of deer browsing, and continued encroachment by sheep . . . which is impacting survival of seedlings'.[22] Indeed, what little natural regeneration I'd spotted on my visit comprised mainly holly bushes, the only tree species spiky enough to withstand the depredations of sheep. If left to get out of control, the evergreen holly will start shading out any oak saplings, making the problem even worse.

To have sheep running riot in this rainforest was not the intention of Johnny Wood's owner, the National Trust, who bought the woodland in the 1960s specifically to preserve it.[23] The entirety of the wood has been fenced off in recent years to try to stop sheep getting in from neighbouring fields. But, as anyone who's ever observed a sheep forcing its way through a small hole in a fence will know, they're persistent buggers with an eye for weakness.

Sure enough, as I rounded the edge of Johnny Wood on my

way to the return bus, I came across a section of stone wall which had collapsed. Some wire-mesh fencing was still in place above the rubble; but snared on its ends were the tell-tale signs of a flock of woolly escapees who'd made their break for freedom. I later spoke to a friend at the National Trust about this breach, who said that sheep weren't *meant* to be in Johnny Wood, but that there was always a risk of them slipping through. Another conservationist I spoke to in the Lakes muttered darkly that the Trust doesn't do enough to check up on its tenant farmers, meaning that the wood was 'full of Herdwick sheep' as a result.

Depressed by this stark example of a rainforest in decline, I was keen to visit another site in the Lake District where the outlook was brighter. So the next day I headed east, to a place that's not only protected but is actively headed for recovery: Wild Haweswater.

The Wild Haweswater project stretches over 3,000 hectares of land surrounding Haweswater reservoir. It's a partnership between the landowner – water company United Utilities – and the Royal Society for the Protection of Birds (RSPB) – who manage the land as farm tenants. At its heart is Naddle Forest, a two-pronged slice of ancient temperate rainforest hugging the sides of the reservoir and the nearby Naddle Beck. I'd been invited to visit by the RSPB's Lee Schofield and David Morris, who showed me around Young Wood and whom I'd worked with previously to try to get the government committed to wilder National Parks.

David's an incredibly knowledgeable ecologist, whose years of battling government inertia on environmental issues has given him a wonderfully cynical sense of humour without wearing down his boundless energy. His Twitter handle – @JFDIecologist – gives a hint at his character: the acronym stands for 'Just Fucking Do It'.

Lee, the RSPB's site manager for Haweswater, affably bears the scars of being a conservationist in a part of Britain that's sometimes hostile to the idea. In his recent book *Wild Fell*, Lee talks about being made to feel unwelcome in the area, because farmers are seen to 'belong', whereas environmentalists are often perceived to be invading townies. 'The prospect of an institution like the RSPB taking on a tenancy . . . was so far from the norm as to be considered close to criminal in some quarters,' he writes. 'At times, however, the discomfort with our presence in the valley spilled over into open hostility.'[24] With his salt-and-pepper hair, sharp nose, flaring eyebrows and penetrating eyes, Lee looks a little like one of the eagles that once frequented Haweswater – until the last one died in 2015.

Making the landscape fit again for eagles is just one of the ambitions of the Wild Haweswater project. Together, we sat around a table in the RSPB's offices at Naddle Farm, sipping tea and studying maps of the area to gauge the lie of the land and decide where we'd later visit. To help local residents and visitors get an idea of the project's scope, Lee had commissioned an artist to paint a series of pictures of Haweswater as it might look in decades to come. A particularly striking bird's-eye view shows the landscape as it is now: the trees huddling mostly in the valleys, the fell tops grazed to within an inch of their lives by flocks of sheep. A second painting shows how it could look by 2040. It's an inspiring vision, one in which the hillsides burst with life once again: trees and scrub regenerating up the cloughs, wood pasture on the uplands lightly grazed by cattle, all creating a rich mosaic supporting curlews, cuckoos – and, with luck, resurgent eagles.[25]

'This is a vision in which farming and nature restoration go hand-in-hand,' Lee told me. 'We'd like to see trees naturally regenerating, feathering up on to the fells, blurring the hard lines that currently separate forest from farmland. But we'll still have grazing livestock – it'll just be very different from how much of the Lake District is managed.'

A few years ago, the RSPB took the unusual step of publishing their accounts for Haweswater, revealing that 'without significant government grants . . . our farm business here would make an enormous loss'.[26] Indeed, the average livestock farmer in the English uplands makes a loss of £10,000 a year, leaving them heavily dependent on taxpayer-funded farm subsidies to make ends meet.[27] Now, post-Brexit, those subsidies are in the process of being reformed, with the Treasury counting every penny. For Lee and his team at Haweswater, something had to give. So they're trying something different: bringing livestock numbers down, and switching from sheep to cattle, as studies show that cattle are much less likely to eat young saplings than sheep are.[28] By applying fewer artificial fertilisers and pesticides, the cost spent on inputs is reduced, and the focus is on outputs of quality meat rather than sheer quantity. As a recent report on the changing economics of sheep farming puts it: 'Less is more.'[29]

This approach makes sense both economically and ecologically in a place like Haweswater – not least because one of its main functions is as a water catchment for a reservoir. John Gorst, United Utilities' jovial catchment manager for the area, was also sipping tea at the table with us. 'The reservoir here is the single largest water supply in the northwest,' he told me. 'It contains about 40 million litres of water, the drinking supply for 2 million people.' Yet maintaining a reliable water supply is getting harder with the escalating climate crisis. Back in 2015, Cumbria was hit by torrential rain and catastrophic flooding; since then, it's experienced droughts in three of the last four years. Water companies have an obvious financial interest in working out how to cope with increasingly extreme weather, as well as a statutory obligation to help prevent flooding. For John, 'making our catchments more resilient to climate change is a no-brainer'.

It turns out that rainforests have an important role to play here. The temperate rainforests that once would have cloaked the hillsides of the Lake District acted as natural flood defences, slowing the flow of rain downhill. Rainwater flashes off a heavily grazed slope far more easily than one covered in vegetation. John pointed at his head, grinning. 'It's like that science experiment in which you pour a jug of water on to someone who's bald, like me, and see how quickly it flows off compared to someone with a full head of hair.' During the 2015 floods, this science experiment played out in real life in the Lake District. David recalled how all the roads surrounding Thirlmere reservoir a few valleys to the west had been blocked by mud and boulders washing off the hills, except for one: a section downhill from a patch of Atlantic rainforest. The tree roots had kept the soil intact and absorbed some of the force of the floodwaters.

Fired up by this talk, I was keen to get out walking to see Haweswater for myself. The four of us pulled on boots and coats and headed up the hill, following the Naddle Beck. Partway along, we paused to look at an area of ash woodland that had been fenced off around the stream. Historically, fencing off bits of land was done for the purposes of 'enclosure' – enclosing a plot so that animals could graze in it without escaping to a neighbour's field. But the fence around this area of woodland and stream was the opposite: an 'exclosure', designed specifically to *exclude* livestock, rather than pen them in. Not only does this prevent overgrazing of the wood, it also stops the sheep from shitting in the water. Sheep poo often contains a microscopic bug called cryptosporidium, which causes diarrhoea in humans but is resistant to the chlorine disinfectant used in water treatment plants. So exclosure fencing along river catchments can be a key part of reducing water pollution downstream.[30]

As we walked further, the woodland edge gave way to grassy moorland and heath, and we talked about what this landscape might have looked like long ago. 'There are loads of ancient woodlands in the Lake District that just don't have any trees left!' said John. I scratched my chin. What he meant was that whilst the trees may have disappeared, clues to their former existence can still be gleaned by looking for the woodland plants growing from the old seed banks. David pointed out patches of wood sorrel at our feet, and talked about the bluebells that would turn parts of the hillside azure in spring: both clear indicators that woodland once clothed these fells.

The stands of bracken that also cover much of Haweswater would be viewed by many farmers as a problem, since livestock tend not to eat it and it shades out grass. But in reality it's another sign that the soils here are probably more suited to trees. 'If you want to solve the problem of bracken in the uplands,' said Lee, 'plant it up with saplings: they'll end up shading out the bracken, and their seeds will be a source of more saplings in turn.' It wouldn't give you back much pasture, but perhaps bracken's appearance is a warning that we shouldn't be trying to eke a crop of grass out of these hills in the first place. In Lee's view, creating a bracken map of the uplands would provide as good a guide as any for where to focus reforestation efforts.[31]

Sheer altitude isn't a factor in preventing trees regenerating in the Lakes, David told me. But you're unlikely to ever get closed-canopy rainforest on the moorland tops, because of the peat that's formed there over the thousands of years since rainforests were prevalent. We passed by a peat bog that David had once measured, finding it to be a staggering thirteen metres deep, and probably accumulating carbon since the last Ice Age, around 10,000 years ago. Peat is a hugely important ecosystem in its own right, and restoring rainforests isn't about playing off one valuable habitat against another. Rather, it's about letting trees return to the areas where they clearly can still thrive – the steeper cloughs, the areas currently dominated by bracken – and creating a much richer patchwork landscape that interweaves peat bogs with rainforests, wood pasture with valley mires.

But rainforests won't regenerate whilst we keep the fells overgrazed, and preventing overgrazing is going to need intervention.

After an hour of trampling through boggy heath, we had now reached another fence. I was taken aback by the stark contrast between the moorland on either side of it, like the before-and-after images we'd looked at earlier over tea. On the approach side, grazing livestock had bitten the grass to the quick; but across the threshold, where grazing was more controlled, bilberry bushes abounded, mounds of sphagnum moss bulged, and rowan saplings were sending up shoots. Resurgent purple heather intermingled with the orange flowers of bog asphodel and the bobbing white heads of cottongrass. The regenerating flora, in turn, were providing a better habitat for ground-nesting birds, like golden plover and lapwing. 'This has all just happened in the past eight years,' David told me.

'Installing fencing on the fells can be incredibly expensive,' said Lee. 'There's often no access roads, so you have to helicopter the materials in.' Many of the uplands in the Lake District are also open access land, as well as commons. And whilst fencing needn't be a barrier to the Right to Roam – the RSPB have put in plenty of gates and stiles – fencing off commons raises the spectre of enclosure, and can provoke resistance from farmers with common grazing rights. The Open Spaces Society, set up in the nineteenth century to resist some of the final Enclosure Acts that enclosed the commons, maintains a position that it is 'likely to oppose any fencing' on common land, although it does sometimes make exceptions for wildlife conservation.[32] As an inveterate trespasser and campaigner for a greater Right to Roam in England, I have no great love for fences. But where their function is to keep out livestock rather than people, I think they have a vital role to play in regenerating our lost rainforests.

United Utilities and the RSPB have succeeded in getting agreement with other graziers on the commons they manage at Haweswater to put in exclosure fences, which now cover large swathes of Mardale Common surrounding the reservoir. But given the costs of installing them, coupled with the fraught politics of commons, they're also looking at other options. Fortunately, it looks like modern technology may have come up with a solution.

As we reached the crest of the hill, Lee scampered off ahead of us, clutching his mobile phone. 'He's gone in search of our herd of Belted Galloway cattle,' explained David. 'He's got them tracked on his smartphone app.'

It's a novel bit of kit called 'Nofence' technology, and it works by fitting a collar containing a GPS tracking device to each cow, connected to an app on your phone. Then comes the clever part. To control where the herd graze, you use the app to draw a 'virtual fence', dividing the land which you want to remain ungrazed from the area where you're happy for the livestock to roam in. When a cow approaches this virtual fence, the collar it's fitted with plays a steadily rising note, warning the animal it's approaching the edge of its range. If it touches the threshold, the collar gives the cow a slight electric shock, just as electric fencing would. The result: controlled grazing, without fences.[33]

Eventually we caught up with Lee, who was watching his herd of gently mooing Belties with something approaching love. The collars they were wearing look a little like Alpine cowbells. 'Soon you'll have to start wearing lederhosen and yodelling,' David chided him. The hills are alive, it seems, with the sound of trees regenerating.

We made our return to the farmstead via the lower reaches of Naddle Forest, along the shores of the reservoir. Tree lungwort bloomed from veteran ashes, accompanied by its rarer cousin, *Lobaria virens*, with its verdigris sheen. Back at Naddle Farm, Lee showed me the old sheep shed that his team are busy converting into a tree nursery: a repurposing that encapsulates in miniature the transition Haweswater is undergoing. 'Where the shit of sheep once was, the trees of the future now grow,' said David, ever the straight-talking Yorkshireman.

It's an ecological transition that also makes economic sense: reducing input costs, fetching a premium for grass-fed beef rather than loss-making mutton, and supplementing incomes with revenues from eco-tourism. Visitors flock to Haweswater already to see the badgers from a new badger hide; imagine the public interest it'll attract when they reintroduce pine martens and eagles to a regenerating rainforest. On top of all that, Wild Haweswater is sequestering carbon, reducing flooding and cleaning up the water supply: all public goods thoroughly deserving of public investment. 'Farming has always responded to the needs of society,' Lee told me. 'Which is exactly what we're continuing to do.'

Considering how many farmers are having to cope with big annual losses, and how the government is struggling to grapple with the climate and biodiversity crises, it all sounded like irrefutable common sense to me. But to understand why things like exclosure fencing, Nofence tech and natural regeneration aren't yet the established orthodoxy everywhere, we need to take a trip back in history.

We've developed something of a mania for planting trees in modern Britain. It's become our new civic religion: each November we now celebrate National Tree Week, and make our votive offerings by digging a hole in the ground and planting a sapling in it. Don't get me wrong – I love planting trees, and it's often a good thing to do. But we seem to have forgotten the fact that trees have been capable of naturally regenerating and spreading for millions of years, if given half a chance.

Tree-planting first became a mass movement in Britain in the 1970s, when a government-sponsored campaign urged the public to 'Plant a Tree in '73'. The immediate spur to action had been the tragic spread of Dutch elm disease, which was decimating the country's elms. But the proposed solution – to plant new trees in their stead – derived from the forestry industry.

The Forestry Commission had, after all, spearheaded the UK's drive to reforest itself since the First World War by planting millions of trees. Planting meant you could choose the species you wanted, ensure they were tidily arranged in straight rows for easy harvesting, and select the best specimens for timber. As an idea, it went back to the seventeenth-century writer John Evelyn, who urged aristocrats to plant trees on their estates for an earlier war effort.[34] You might say that planting trees is a habit with deep roots in the British psyche.

Alongside this strengthening belief in the need for tree-planting came a declining faith in the power of trees to naturally regenerate and spread. Natural regeneration was 'rarely practised in Britain', reported the forester Lord Bolton in 1956.[35] 'Many foresters believe that there are biological and climatic reasons why good regeneration cannot be achieved in Britain,' found a Forestry Commission briefing in 1995. But in reality, it admitted, this had little to do with science, and more to do with the fact that 'most foresters . . . have had little training in or experience of natural regeneration and consequently lack confidence in the application of the system'.[36]

Oak trees, in particular, were perceived to be feeble and incapable of replenishing themselves. For many decades, foresters debated why oaks in modern Britain did not seem to be regenerating and spreading. The forester John Nisbet wrote in 1894: 'Many . . . oak woods are now found difficult to regenerate naturally.'[37] A century later, foresters were still warning that 'natural regeneration of oak is usually accomplished in Britain with difficulty, when it can be accomplished at all'.[38]

Some pointed to the fact that oaks don't reliably produce acorns on an annual basis, but rather have 'mast years' in five-to ten-year cycles: a boom in acorns followed by a bust. If the mast year fails to result in lots of new oak seedlings, it can be a big setback for regeneration. In Lord Bolton's view, 'Oak regenerates sporadically . . . I would not care to rely on it for the perpetuation of a wood, and would rather look on self-sown oak seedlings as a gift from God.'[39] Others blamed oak's apparent inability to reproduce on the intolerance of oak saplings for shade: and it's true that acorns that germinate beneath the shade of the mother oak are often destined to fail. Oak mildew, rabbits and caterpillars were amongst the other reasons cited by forestry experts as to why natural regeneration just wasn't viable.[40]

All of these factors are challenges that oaks, in common with other trees, have to cope with in order to survive and thrive. But if these reasons were all so overwhelming, why does Britain have woods at all? And why did they once cover far more of the country than today, long before people were around to plant them?

What the critics of natural regeneration had forgotten was that different natural processes also help oaks and other trees to spread, rather than merely hindering them. The real problem was that we – humanity – had upset the balance. Firstly, modern industrial agriculture had raised the numbers of sheep and cattle in Britain to unnaturally high levels, meaning over-grazing was destroying huge numbers of saplings. The breeding of rabbits for fur in the eighteenth century and deer for sport in the nineteenth had also massively bolstered the number of nibbling and browsing animals in British woods. Secondly, farmers had cleared much of the thorny scrub that naturally protects young seedlings from being eaten when they're starting to grow. And thirdly, in our anthropocentric assumption that humans are the only ones who plant trees, we forgot the role played by other animals, like jays and squirrels, in helping transport seeds far from their mother tree to areas of fresh ground with plenty of light.

Eventually, forestry experts began to recognise that over-grazing was critical in preventing the natural regeneration of oaks and other species. A 1991 review by the Forestry Commission admitted that 'heavy grazing . . . is thought by many to be the most important single factor limiting regeneration in many woods'.[41]

Whilst decades of scrub clearance, driven by perverse agricultural subsidies and a mania to 'tidy up' the countryside, has led to the widespread removal of this crucial aid to natural regeneration, the value of thorny scrub is at last starting to be rediscovered. This is thanks in large part to the work done by the Knepp Estate rewilding project in Sussex. As its co-owner, the appositely named Isabella Tree, reminds us: 'The thorn is the mother of the oak.'[42] This old piece of wisdom dates back at least as far as 1613, when the writer Arthur Standish stated that 'in all ages, bushes have beene . . . the mother and nurse of trees'.[43] I've witnessed this truth myself many times on Dartmoor, where oaks and rowans can sometimes be seen peeping out of the tops of gorse bushes, having evaded the sheep.

What's more, fresh research is confirming that we humans have some way to go to measure up to nature's most prolific tree-planter, the humble jay. With their electric-blue wing feathers, jays are instantly recognisable, but their role in natural seed dispersal has perhaps been under-recognised. Jays are particularly important in helping oak trees spread: during mast years, they collect hundreds of acorns, taking them far from where they fall and burying them for safekeeping. (Their scientific name, *Garrulus glandarius*, means 'babbler of acorns'.[44]) But they tend to forget where every acorn has been buried, so some will germinate into saplings. And by carrying acorns hundreds of metres from the mother oak into surrounding fields, jays help oak saplings escape from under the shaded canopies of their ancestors, emerging instead into bright sunlight. A recent study of natural regeneration taking place in a field bordering an old oak woodland has shown that half the trees in the new wood were planted by jays.[45]

In fact, we likely have jays to thank for Britain having

oakwoods at all. After the last Ice Age, the land-bridge joining Britain to continental Europe was soon submerged by melting ice. Trees and vegetation colonised the lands exposed as the glaciers retreated. But how did the supposedly slow-spreading oak make it across the Channel in time? The ecologist Clifton Bain suggests jays provide an answer. Had oaks been forced to disperse their acorns through gravity alone, he argues, it would have taken 100,000 years for them to spread from the continent up to Scotland – missing the boat as the meltwaters advanced. But the pollen record shows a more rapid march of oaks northwards of about 500 metres per year. 'Perhaps, had it not been for the jay assisting the progression of the forests, oak may not have reached our island before rising sea levels cut us off from mainland Europe,' he says.[46]

There are still plenty of jays in our countryside today, yet far more sheep and far less thorny scrub than there used to be. Encouraging more hawthorn, gorse and brambles to grow on the edges of woodlands and shelter new saplings is surely the medium-term solution. But, in the meantime, there's also a more immediate way to deter sheep from grazing in the woods: put up fences around them.

The breakthrough study in understanding the value of exclosure fencing came from an ecologist called Donald Pigott. His research focused on Yarncliff Wood, in Padley Gorge in the Peak District, which just so happens to be a candidate for temperate rainforest owing to the gorge's damp, sheltered microclimate. When Yarncliff Wood was first surveyed in the 1950s, no oak tree could be found that was less than a hundred years old. A proposal to fence off the whole wood to allow regeneration to occur was rejected by its owner, the National Trust, because it was believed at the time that the site's altitude and poor-quality soil were the main causes of the lack of young trees. So Pigott decided to set up a long-running experiment. In 1955, he persuaded the Trust to create a smaller exclosure within the wood, and allow him to observe the changes that took place.

Pigott must have had the patience of a saint, because he continued monitoring the exclosure plot for almost thirty years. At last, in 1983, he felt confident enough in the results to publish his findings. In the woods outside the exclosure, sheep had continued to keep the forest floor bare of saplings and pretty much every other plant. But where the sheep had been excluded by the fencing, the changes had been transformational. A vibrant under-storey of bilberries, ferns and woodrush bloomed, and oak saplings had regenerated, reaching heights of five feet; some were even growing underneath the shade of other oaks, despite many at the time believing such regeneration could never occur. 'The experiment shows convincingly that sheep were and still are responsible for failure of regeneration of trees in Yarncliff Wood,' concluded Pigott triumphantly, 'and that, when sheep are excluded, all the species of tree originally present can regenerate from seed.'[47]

The results were so compelling that the National Trust reversed their policy of allowing sheep to roam freely in their ancient woods, and began – gradually, at least – to put up more exclosure fencing. One woodland owned by the Trust to benefit from this change of heart has been Keskadale Wood in the western Lake District. High up on the slopes of Ard Crags, it attracted the attention of the pioneering botanist Arthur Tansley in the 1940s, who noted its abundance of epiphytic mosses and liver-worts, but commented: 'Oak seedlings are rare . . . It is not clear that the woods regenerate.'[48] Later surveys by the National Trust agreed. Maurice Pankhurst, the Trust's woodland ranger in the Lakes, noted a 'complete absence of natural regeneration', and warned that 'should the current period of grazing continue, then it is difficult to predict how much longer it can survive'.[49] In 2006, the National Trust put a wide exclosure fence around Keskadale Wood, allowing it a respite from grazing for perhaps the first time in centuries.

Other landowners, too, have cottoned on to the idea of giving our fragments of temperate rainforest a breather from constant grazing. At Piles Copse, one of the three upland oakwoods that survive on Dartmoor, the owner John Howell has put in a number of small exclosures around the edges of the wood.[50] Some time after my trips to Wild Haweswater and Johnny Wood, I visited Piles Copse and saw for myself the positive difference made by the fencing: the young oak and rowan saplings and rejuvenating ground flora contrasted greatly with the aged trees and lack of understorey in the rest of the woodland. Significantly, the exclosures here are only in small blocks on the very edge of the wood, rather than encompassing the whole of it. This means the older, established rainforest retains its open character, rather like a wood pasture, and full public access to it is maintained, whilst also allowing a new generation of trees to develop on its perimeters.

Another Dartmoor landowner who's experimented with exclosure fencing is Leonard Hurrell. I met Leonard and his sons Robert and Andrew on a rainy August day (this being the rainforest zone, after all) to look around the land they own on Ugborough Moor, in southern Dartmoor. Being treated to tea and scones in Leonard's amazing old house on the edge of the moor was a bit like stepping back in time: pleasantly musty smelling, filled with huge numbers of books on natural history and hung with wonderful chalk sketches of otters and birds drawn by his father, the renowned naturalist Henry Hurrell.[51] A family of keen conservationists, the Hurrells have tried for years to encourage the natural regeneration of trees on the moorland they own. But there's a slight catch: most of Ugborough Moor is a common, and getting all of the commoners to agree to put in even small areas of exclosure fencing has proven difficult.

Despite this, the Hurrells have persevered, obtaining reluctant agreement from the commoners to put in two livestock exclosures around a steep-sided boggy brook, which they've re-vegetated with a mixture of tree-planting and self-seeding. Where trees already exist to provide a viable seed source for new saplings, it surely makes sense to let natural regeneration do the job. But in parts of our uplands, where centuries of overgrazing have removed all tree cover, planting trees absolutely has a place in getting things moving again. It's seldom easy, though. Andrew told me how he had planted many saplings within stands of gorse to protect them – 'a prickly business', as he said – only to see them go up in flames. This was no accident, but a result of the annual 'swaling' season, when the commoners burn any gorse and scrub that's grown up on the common, to produce better pasture for livestock.

Exclosures are clearly a key tool in helping bring back our lost rainforests, but what about simply reducing sheep numbers overall? In 2001, the Hurrells were witnesses to the impact of suddenly removing sheep and cattle from the moor as a result of the devastating foot-and-mouth crisis. Andrew recalled walking to the top of Ugborough Beacon after the end of the crisis and finding oak saplings growing there. He had never seen any growing there previously, and never has since. The fact that the Beacon lies about 380 metres above sea level puts paid to the idea that Dartmoor's uplands are too high up and exposed to support trees. It had taken less than a year of sheep being absent from the moor for natural regeneration to begin afresh.

Clearly, no one wants a repeat of foot-and-mouth. But with farm subsidies changing, and Brits cutting their meat-eating by 17 per cent in the past decade, the writing may already be on the wall for densely stocked upland sheep farms.[52] The challenge is to make sure farmers and landowners are supported in making such a transition. Even in parts of the Lake District, the sheep are starting to reduce in numbers. The Lakeland shepherd James Rebanks has written about the recent changes in his book *English Pastoral*. It reads a little like the shifting landscape described in

Wordsworth's *Guide*, only in reverse. 'I see ancient oak woodland above us trying to regenerate,' says Rebanks. 'Little mountain ash trees are sprouting up all over the wilding fell, trying to beat the deer . . . The valley has become much shaggier and wilder than it ever was in my childhood, with far fewer sheep dotted around. Some of my neighbours are confused or angry about that, while others are adapting, keeping more cattle or finding other ways to earn a living from the land.'[53] This is not a landscape preserved in aspic, but one in which natural processes are gradually returning.

Unfortunately, stopping sheep from eating our temperate rainforests is only part of the battle. Simply fencing off our surviving rainforest fragments would help the natural regeneration of saplings, but it also risks creating a different sort of artificial environment. The overgrazing suffered by our rainforests for centuries is unnatural; but so is a woodland bereft of all animals.

There's a particularly delicate balance to be struck around fostering the conditions in which rare rainforest lichens and bryophytes can thrive. These epiphytes are, after all, what define our rainforests and make them of international significance.

As we've learned, many rainforest lichens are very particular about needing just the right combination of sunlight, moisture and shelter to thrive – the 'Goldilocks zone' in which they flourish. A regenerating understorey of vegetation has the potential to swamp lichens growing on the trunks of trees, blocking out the sunlight they need. Some lichenologists argue that simply banning all livestock from our rainforests could end up having a detrimental impact if overgrazing is replaced with overshading. As the lichens expert April Windle says, 'Lichens need light in order to photosynthesise, and woodland habitats with higher and more varied light levels have a larger diversity of lichens.'[54]

At the same time, there are bryologists who warn that opening up too much of a forest canopy risks drying out the moist, shade-loving mosses and liverworts that also make our rainforests what they are. The ecologist Derek Ratcliffe has noted: 'The majority of our Atlantic bryophytes show some need for a moist atmosphere, and are thus associated with conditions which reduce evaporation, such as the shade of trees in canopy, or sunless aspects . . . of outcropping rocks.'[55]

This sets up quite a quandary. If our rainforest fragments are always prevented from regenerating, they'll eventually die. By remaining as isolated remnants, cut off from one another, they also lack the connectedness that would allow the other species they support to spread more easily across the landscape. This goes for lichens as much as for other species, whose soredia and fungal spores seldom travel very far, even relying on raindrops to carry them from branch to branch. Reconnecting old-growth rainforests by letting them self-seed and spread on to neighbouring land is a critical step in breaking this cycle of decline. But there's little point regenerating the trees in our rainforests only to lose the lichens and epiphytes that make them special.

It may, however, be possible to find the sweet spot between these two extremes. Something else that is missing from our temperate rainforests is natural levels of disturbance. Once trees reach a certain age and size, they become less susceptible to grazing pressures from livestock and to browsing by deer. At this stage, lower stocking densities, mimicking natural grazing levels, is often preferable to zero grazing. 'It's really complicated,' says Luke Barley, woodlands advisor at the National Trust, when I speak to him on the phone. 'We may never find the ideal stocking level that creates the perfect conditions in any one wood in perpetuity. It's probably more natural to have *pulses* of grazing, where the level varies over years and decades.'

The bigger the woodlands, the easier it may become to resolve these trade-offs. Greater scale brings with it the chance for greater variety: you prioritise some areas for regenerating young trees, and manage other areas to optimise conditions for lichens.

It's hard to recreate the conditions of the original Wildwood when all that you have left are postage-stamp-sized woods. But by increasing the size of our rainforests, conservationists won't have to fight over the scraps that currently remain.

By swapping sheep for cattle, you also replace a non-native animal with one closer to the wild aurochs that would once have roamed our woods thousands of years ago: one whose incisors are less able to nibble saplings down to the ground. Those herbivores then have a key role to play in creating useful forms of disturbance in a forest: trampling down bracken, browsing resurgent holly bushes, knocking off old branches, creating glades that let the sunlight in. Tools like exclosure fences and Nofence technology make it easier to control grazing levels and vary it over time and by location – you can decide if a wood needs to be given a break, or could cope with a little grazing.

In the past, our temperate rainforests would have been alive with numerous other species of animals, many now extinct in Britain. Reintroducing these missing species – or close proxies to them – can help restore other natural processes and create dynamic new patterns of disturbance. Bison, for example, are an even closer proxy to wild aurochs than cattle. They're huge creatures, capable of smashing through branches and pushing over old trees with their powerful horns. This might seem destructive, but in reality it's just restoring some of the natural mess and chaos that's long been missing from our woodlands. At the ancient woodlands of the Blean in Kent, the local Wildlife Trust has plans to reintroduce bison from eastern Europe in 2022. These 'ecosystem engineers', it hopes, will help create glades in the woods, letting in more light and generating a richer mosaic of habitats in which lots of species can thrive.[56]

Beavers, now making a joyous comeback in the UK after they were hunted to extinction for their fur 400 years ago, are well known for their incredible ability to build woody dams and hold back floodwaters. The habitat they've created at the Cornwall Beaver Project, where they were reintroduced some years ago by farmer Chris Jones, is nothing short of astonishing: a series of interlocking pools and swampy woods, heady with the smell of water mint and primeval in feel. But beavers don't just replumb a river: their fondness for gnawing down trees also clears space in the canopy, letting in light. Still, there are trade-offs here, too: beavers have little respect for whether a tree is young or old, and have been known to chew through ancient oaks in Scotland supporting rare species of internationally important lichens.[57] Ultimately, however, the more rainforest habitat we bring back, the more it ought to be able to absorb this sort of loss.

Wild boar and native breeds of pigs could also have important parts to play in rejuvenating rainforests. They're great at bashing down bracken and brambles, helping prevent such scrub from forming an impenetrable monoculture whilst still retaining clumps that protect young saplings. And with their tusks and love of snouting about in the earth for roots, insects and fungi, wild boar are excellent at creating ground disturbance. They're particularly good at breaking up thick mats of grass, gouging out chunks of turf in a field to reveal bare soil: something I've witnessed in the Forest of Dean, where a feral population of boar have merrily ploughed up grass verges everywhere. Seeds that fall on this fertile ground, whether from trees or other plants, find it easier to put down roots – speeding up the process of natural regeneration.

What about restoring the most controversial natural process of them all: predation? After all, some of the threats faced by our rainforests would be better kept in check with some reintroduced predators to help. England's plague of grey squirrels, introduced from North America by a foolish Victorian aristocrat, has almost exterminated our native red squirrel by spreading a deadly pox – and also caused huge damage to trees, as greys strip the bark off trunks to use as nesting material. But grey squirrels are preyed upon by pine martens, a gloriously slinky mustelid related to weasels and stoats. The problem is, pine martens are extremely rare in England, having been driven out over a century ago by deforestation and gamekeepers. Bringing them back would also help our rainforests recover, and boost the chances of red squirrels returning from the brink.[58]

And whilst well-maintained exclosure fences might keep out sheep, the same can't be said of deer, who will vault all but the highest deer fencing. That means conservationists are forced to reduce deer numbers at source: either by culling – or bringing back more of their natural predators. Both lynx and wolves prey upon deer, and would have a great impact on controlling their browsing of trees: not just through the sheer number they would kill for food, but through the 'landscape of fear' they'd generate, deterring deer from venturing into large swathes of forest.

Of course, to get to this stage, we Brits will need to overcome our own fear of the wolf, hardwired into us by Grimm's fairy-tales and a thousand years of keeping the wolf from the door. But wolves can now be found coexisting alongside people even in the Netherlands, one of the most densely populated countries on earth. The inevitable impact that wolves have upon sheep is simply accepted in much of Europe, where conservationists and hunting organisations from France to Romania have long chipped in to cover insurance premiums for farmers to compensate them for herd protection measures and loss of livestock.[59] But politically, it will still be a brave government that oversees the first reintroductions of wild carnivores to Britain, particularly the first release of a wolf into the 'Republic of Shepherds'. The problem we should be getting on and resolving in the meantime is that our rainforest habitats remain too small and too fragmented to realistically support apex predators.

We are back to where we started, dreaming with the Romantic

poets of changed landscapes and lost species. This time, however, we're not just dreaming of the past, but of how things might yet be different in the future.

We may be waiting a while to see the fulfilment of Wordsworth's reverie for a primeval Lake District of eagles and wolves. But as for his vision of 'the universal sylvan appearance the country formerly had'? Listen carefully, and you can already hear the saplings unfurling their leaves.

7

'Nobody Cares for the Woods Anymore'

As soon as I pushed open the rusty iron gate and stooped under the tendrils of ivy, I knew I'd come to the right house. The dark, winding garden path was almost overgrown with plants; a Green Man sculpture peered out from the undergrowth next to the porch. With a mixture of excitement and trepidation, I reached up to knock on the front door and saw that the door-knocker was in the shape of a pixie.

I'd come to visit one of my childhood heroes, the J. R. R. Tolkien illustrator and fantasy artist Alan Lee. As a teenager, obsessed with playing *Dungeons and Dragons* and watching fantasy films like *Labyrinth* and *The Dark Crystal* – I still have the papier-mâché head I once made of the vulture-like Skeksis – I'd pored over Alan's dream-like watercolours and drawings with awe.

With aspirations at the time of one day becoming an illustrator, I remember spending hours trying to copy one of Alan's drawings during A-Level art classes: an extraordinary pencil rendition of Merlin as a shaman of the woods, his tattered robes and cloak of crow-feathers merging with the branches of a tree and the rushing waters of a stream. There's plenty of awful fantasy art out there, all airbrushed armour and damsels in distress. But Alan Lee stands apart: a master draughtsman, whose lonely figures exist in vast, moody landscapes that are characters in themselves.

The door opened, and a face appeared: silver hair, twinkling blue eyes, impressively bushy eyebrows. 'Come in!' Alan greeted me.

Alan's house and studio were exactly as I'd hoped they would look: like the fantastically cluttered interior of Merlin's cottage in his cover illustration of *The Sword in the Stone*. I felt a little like the wide-eyed young Wart being shown around by the master of magic. Animal skulls, clay heads, willow sculptures, endless bottles of ink; hundreds of paintbrushes and pencils in jars, photo-montages of trees, tubes of paint. The walls practically leaned in under the weight of books, covering everything from Celtic mythology and Pre-Raphaelite art to – of course – all the works of J. R. R. Tolkien.

After illustrating a special edition of *The Lord of the Rings*, Alan was head-hunted by the film director Peter Jackson to become one of the lead concept artists for his movie adaptations, for which he won an Oscar for Best Art Direction. The golden statuette gleamed proudly on a shelf, near a figurine of Treebeard, the shepherd of the woods in the trilogy. The fanboy in me couldn't resist asking for a photo with Alan in his studio, gabbling an apology for being cheesy. 'We could make it even more cheesy if you like,' said Alan, and he fetched the staff – yes, *the* staff – wielded by Gandalf in the film adaptation of *The Fellowship of the Ring*. It had been given to him after finishing work on the movies. 'Here's where Gandalf's pipe slots in,' he showed me, 'and here's the pouch of flints and tobacco.'

But I hadn't just come here to swoon; I also wanted to ask Alan about the woods that feature in his art. In a collection of his illustrations, I'd found a pencil sketch of an ancient tree covered in epiphytic ferns and girded with mosses and lichens. The caption related that he had drawn it from life near

Dartmoor: 'The moorland of Devon is so powerful an attraction that he constantly returns to it as a source of inspiration.'[1] It turned out that Alan had moved to live near Dartmoor's wooded valleys in the 1970s. 'I was just amazed at how lush it all was,' he once recounted to an interviewer.[2] Here, he found 'as many moss-covered boulders, foaming rivers and twisted trees as my heart could ever wish for'.[3]

Many of Alan Lee's illustrations have been inspired by the rainforests of the Westcountry. When I first contacted him to request a meeting, Alan confirmed that he was 'a big fan of damp forests'. In his studio, he showed me the gorgeous watercolour cover he had done for an illustrated edition of *The Mabinogion*, the book of Celtic legends that – as we saw in chapter 5 – are set amongst Wales' rainforests. At its centre is a huge tree overhanging a river, ivy encircling it like a coiled serpent; the base of its trunk is darkened with moss and mottled with bracket fungus. A column of warriors on horseback ride behind it, almost swallowed by the tangle of surrounding forest. As in much of Alan's art, nature is foregrounded, with humanity's vainglorious conflicts playing out on a much broader and older canvas.

You can also see the influence of Britain's rainforests in Alan's depictions of Tolkien's Middle-Earth. His paintings of Fangorn Forest – the great wood through which the hobbits Merry and Pippin journey in *The Two Towers* – show gnarled boughs dripping with beard lichens and sprouting polypody ferns.[4] Indeed, Tolkien himself describes the forest in this way. In the book, Pippin stares at the trees and exclaims: 'Look at all those weeping, trailing, beards and whiskers of lichen!'[5] The hobbit could have been describing any number of woods on Dartmoor's edge.

But it is Alan's portrayal of the Ents – the huge, dryad-like beings that keep guard over Fangorn Forest – which is most clearly inspired by our rainforests. Tolkien's own description of Treebeard, oldest and wisest of the Ents, brings to mind a lichen-clad rainforest tree: 'the long face was covered with a sweeping grey beard, bushy, almost twiggy at the roots, thin and mossy at the ends'.[6] And in the *Lord of the Rings* films, the Ents literally bear the imprint of Dartmoor's rainforest trees. In a BBC interview, Alan recounted that during the design process, he 'took close-up photographs of bark of trees on Dartmoor, and they were put into the computer. The photographs were used to create the texture of the ents . . . There are some wonderful trees around the edges of the moor, and around Castle Drogo.'[7] So, after looking around Alan's studio, that's where I suggested we go for a walk.

The ancient oakwoods around Castle Drogo on Dartmoor's northeast fringe are magnificent. They cluster round the steep-sided valley of the River Teign, continuing for six miles through an area called Fingle Woods to the east. Alan took me on one of his favourite walks through the Whiddon deer park, an old expanse of wood pasture to the south of Castle Drogo with gnarled and characterful trees. Many were festooned with polypody ferns and furry with lichens; I found some *Sticta* growing on one. A veteran beech, its trunk covered with fissures and knot-holes like eyes, peered down at us with a quizzical expression. Whilst Alan's drawings of trees are usually an amalgam of observation and invention, it's not hard to see where he gets some of his inspiration.

I asked Alan what drew him to woods. 'It goes back to my childhood,' he said. 'I grew up in Uxbridge in west London on an estate. In those days, kids were allowed to roam wherever they liked. We used to go to the canal, pilot one of the old abandoned narrow boats moored up there, and go to explore some nearby woods.' He paused, lost in memory. 'Most of them have now been chopped down and grubbed up, of course.'

We talked about Tolkien's own affinity for woods. Whilst there's scant evidence that Tolkien visited Dartmoor's rainforests, there's plenty of proof that he admired trees.[8] They crop up repeatedly in his works. 'I am (obviously) much in love with plants and above all trees, and always have been,' Tolkien once wrote in a letter. 'I find human maltreatment of them as hard to bear as some find ill-treatment of animals.'[9]

Many commentators have interpreted Tolkien's books as a reaction against industrialisation and the modern war machine: the rural Arcadia of the Shire versus the industrial wasteland of Mordor. The historian Meredith Veldman argues that Tolkien was an early environmentalist, part of a mid-twentieth-century Romantic revolt against the accelerating destruction of the English countryside: 'Decades ahead of the Greens, he denounced the exaltation of mechanisation and . . . the degradation of the natural environment'.[10]

The most obvious eco-warrior in *The Lord of the Rings*, of course, is Treebeard. When he discovers that part of Fangorn Forest has been destroyed by the evil wizard Saruman, in order to fuel the fires of industry and forge metal weapons for his Orc army, Treebeard is enraged and raises up an army of Ents to attack Saruman's fortress. The march of the Ents recalls the animated trees of *The Mabinogion*, and of Birnam Wood coming to Dunsinane in Shakespeare's *Macbeth*. But it also has an added resonance to modern ears: the protest marches of environmentalists opposing the destruction of the rainforests – many of whom read Tolkien and found it chimed with their counter-cultural values. To Veldman, 'Tolkien depicted the trees not as mere lumber but as living beings whose murder causes sorrow throughout the ancient forest of Fangorn.'[11]

Having crossed the River Teign and climbed up Hunter's Tor to its north, Alan and I paused to catch our breath and soak in the spectacular views. The Atlantic oakwoods lining the valley stretched away beneath us, burnished bronze in the low autumnal light. But in the distance, down towards Fingle Woods, the dark spearheads of conifers loomed, like an army of approaching Orcs. Had Treebeard stood guard over the woods of the Teign valley, rather than Fangorn Forest, he would have been similarly enraged by the destruction wrought on it during the middle decades of the twentieth century. For at the same time that Tolkien was writing *The Lord of the Rings*, a rainforest was being cut down on Dartmoor.

For nearly a century, the Dartington Estate in Devon has been renowned as a hub for counter-cultural thinking. Located just outside Totnes – the hippy capital of the West Country – the fourteenth-century stone walls of Dartington Hall today play host to hundreds of students studying ecological economics at Schumacher College, and learning about agro-forestry on its organic farms. But since moving to Totnes in 2020, I've discovered that the history of the Dartington Estate contains a dark secret. This is the story of the bohemians who destroyed a rainforest.

The Dartington Estate was acquired in 1925 by Leonard and Dorothy Elmhirst. Dorothy was a wealthy American heiress; Leonard, the son of a Yorkshire parson. Together, they had a vision to 'take over an old estate in a rural area . . . [and] transform it into an active centre of life'.[12] They chose Dartington, an 800-acre estate whose crumbling Great Hall – built around the time of the Peasants' Revolt – had once played host to dukes and earls, but which had now fallen into such disrepair that the roof had caved in.[13] Dorothy's passion was for education and music: she was a progressive reformer who set up the Dartington College of the Arts, and hosted eminent artists and bohemians from the Bengali poet Rabindranath Tagore to the British potter Bernard Leach. Visiting intellectuals would debate world affairs with their avant-garde hosts, lounge in armchairs reading the *New Statesman*, and watch Ibsen plays staged by the Russian director Michael Chekhov.[14] Leonard, meanwhile, was interested in how a programme of 'rural reconstruction' could lift England out of a decades-long agricultural depression caused by cheap grain imports and, more pressingly, in how he could make Dartington's finances self-sustaining.

To this end, Leonard began making investments in forestry. In 1926, he set up an estate woodlands department, which later became a company, Dartington Woodlands Limited. At first, the forestry activities were modest: some old oaks in Dartington's grounds were felled for timber beams to rebuild the roof of the Great Hall. But Leonard's ambitions went beyond such practicalities. He wanted to experiment with the latest techniques in modern silviculture – the science of cultivating trees – and build a forestry enterprise that would inspire others to follow suit. He hired Wilfred Hiley, a lecturer in forest economics who, like Tolkien, was an Oxford don, and set about acquiring 2,000 acres of woods across Dartmoor's eastern fringes.

The acquisitions included a number of woods that we would now recognise as being temperate rainforests. The 340-acre King's Wood near Buckfastleigh was bought by the Dartington Estate in 1928, whilst roughly 1,000 acres were purchased in the Teign valley, with Fingle Woods being the core. Both lie within the oceanic zone where rainforests thrive, and both were ancient woodlands.[15] This means they were at least 400 years old; dating back to when Shakespeare was writing about Birnam Wood marching on Dunsinane in *Macbeth*, in fact. We know they were predominantly oakwoods, because Hiley recorded this in his later book, *A Forestry Venture*. King's Wood was 'covered in oak coppice and scrub', whilst Fingle and the Teign valley woods were 'all covered in oak coppice' – coppicing being a traditional form of woodland management in which tree stems are periodically cut to promote the regrowth of fresh shoots.[16] In an unpublished ecological survey of woodlands passed to me by Dartmoor National Park Authority, King's Wood is recorded as originally being 'type W17' woodland: the upland oakwood typical of Atlantic rainforests.[17] Unfortunately, it was not to remain this way for long.

For centuries, oak trees had been revered as a symbol of England, cherished for their robustness and strength. Oaks were, as we have seen, sacred to the Celts, and Green Men sprouting oak leaves began appearing in church carvings during the Middle Ages. Edmund Spenser, in his Elizabethan epic *The Faerie Queene*, spoke of 'the builder Oak, sole king of forests all'.[18] The forester John Evelyn praised 'the excellency of the timber' of oaks, which provided the hulls for ships in the Royal Navy from the time of Henry VIII through to the Napoleonic Wars.[19] But the oaks of Dartmoor's temperate rainforests were stunted and gnarled through age and weather. They had been coppiced for centuries for charcoal, and their bark used for tanning leather, but their twisted trunks offered next to no timber. As a result, they held little value for Hiley, who oversaw Dartington's forestry operations from the 1930s until the late 1950s.

Hiley disparaged the Atlantic oakwoods as 'useless coppice' and mere 'scrub' in need of clearing.[20] Other forestry experts at the time agreed. The 1956 Collins New Naturalist guide *Trees, Woods and Man*, written by the forester H. L. Edlin, includes a photo of the Teign valley woodlands labelled 'coppiced oak scrub'.[21] The first official guide to Dartmoor National Park, issued in 1957, described the oakwoods of the Teign and other Dartmoor river valleys, noting: 'To the forester they must be anathema, for the shallow soil restricts the timber severely and it is exceptional to find any trees of respectable size and girth . . . the woods are no longer economic.'[22] The aristocratic landowner Lord Bolton wrote about his experiences of growing trees in a 1956 book, *Profitable Forestry*, and had this to say about the oak: 'As this is our national tree it will always be planted for sentimental reasons; but its place lies in parks and ornamental woods rather than in commercial forests.'[23]

What modern forestry demanded instead of these slow-growing oaks was fast-growing conifers – and lots of them.

So the Dartington Estate cut down these ancient rainforests

and replaced them with conifers. Whilst hardwoods like oak can take over a century to reach maturity, conifers are ready to harvest in around forty years.[24] King's Wood was 'coniferised' first, with 300 acres of it destroyed for plantations of exotic larch, Douglas fir and Sitka spruce in the 1930s. Fingle Woods survived for longer, but they too succumbed to the axe after the Second World War. In place of the old oaks, young conifers were planted and harvested for timber, pit props, firewood and kindling.[25] Reading through Dartington's forestry archives, with their neatly typed accounts and profit-and-loss tables, I was struck by how an entire ecosystem was quickly reduced to a mere resource. Trees are referred to as a 'crop'; untouched habitats as 'unproductive'; areas where scrub and saplings are naturally regenerating as needing an 'urgent clean'.[26] Tolkien's evil wizard Saruman was described by Treebeard as having a 'mind of metal and wheels'; Hiley, it seems, had a mind of spreadsheets and profit margins.[27]

Yet Hiley saw such wanton destruction as a 'considerable achievement'.[28] During the early days of Dartington's forestry enterprise, he wrote, 'it did not seem likely that we could ever clear and replant the vast area of oak coppice and scrub' at Fingle Woods. But after the war, clearance proceeded apace, and 'the whole valley came to have a new look. There was then no widespread aesthetic repugnance to conifers, such as has grown up since,' sniffed Hiley reproachfully.[29] Incredibly, even the carbon-rich peat soils underlying the oak trees in Fingle Woods were dug up and sold as fertiliser. Since the woods had been used 'possibly for hundreds of years as oak coppice', Hiley wrote, a peaty leafmould had built up; in some places, 'the peat may be anything up to a foot in thickness . . . This peat was a serious problem in replanting.' He declared himself 'very pleased' when a contractor offered to pay '£5 an acre for the privilege of removing it'.[30]

Some might excuse such desecration as simply resulting from

Hiley and the Elmhirsts 'not knowing any better'. Clearly, few people in the mid-twentieth century were aware of climate change, or the imminent collapse in species. But the Elmhirsts were ahead of their time in many ways, and liked to think of themselves as environmentalists. Leonard listed his hobby in *Who's Who* as 'care of trees', and his friend Michael Young records that he would 'sit for hours on a shooting-stick in a clearing, talking seriously to other men in plus-fours, all of them craning their necks to look up at the "canopy", to use one of the favourite words of both Leonard and Hiley'.[31] This care for trees, however, doesn't appear to have included not cutting them down.

There was also an awareness, even in the 1950s, that Atlantic woodlands were special places. Yarner Wood, an ancient oakwood a few miles to the south of the Teign valley, was designated a National Nature Reserve in 1952 in recognition of the panoply of wildlife it supported. The forester H. L. Edlin wrote in 1956:

> *In the rainy west, quite a variety of plants find enough moisture on tree trunks to grow thereon; these epiphytes include several ferns, mosses and lichens; while occasionally, in the fork of some forest giant, sapling hollies or rowans may spring up. Such aerial plants draw no sustenance from the soil, and emphasise the fact that forest rainwater, bearing atmospheric dust and the leachings from tree leaves, is itself sufficiently rich in mineral food to support plant life.*[32]

Such dawning awareness, however, seldom led to the region's rainforests getting greater protection.

Cutting down ancient woodland and replacing it with non-native conifers isn't just bad news for the old trees. It's fatal to the whole rainforest ecosystem. Oak trees in Britain support an amazing 284 species of insects; spruce trees, just thirty-seven.[33] The dark, serried ranks of evergreen pines favoured by the Dartington foresters shaded out the understorey plants and killed the remaining mosses and lichens. As the ecologist Derek Ratcliffe warned in 1968, 'conifers are grown in such dense canopy that the intense shade soon eliminates the majority of bryophytes. The prevailing trends in commercial forestry, of planting conifers or converting deciduous to coniferous forest, are thus inimical to the survival of most Atlantic bryophytes.'[34]

Even if the Elmhirsts were ignorant of the ecology of their rainforests, their decision to fell them still seems senseless. There was certainly a national need for timber during the Second World War. Yet the war ironically prolonged the survival of the estate's remaining oak coppices by generating demand for charcoal, which was used for the filters in gas masks. Leonard's investment choices beg the question: why grub up ancient woodland to plant new trees? Why not establish plantations on marginal farmland, as the Forestry Commission had been doing since its foundation in 1919?

It wasn't that the Elmhirsts lacked noble ideals, or had no appreciation of aesthetics; quite the opposite. But ultimately, it came down to cold, hard economics. Old woods on steep land and poor soils were cheap to buy. Hiley expressed it most truthfully in a draft synopsis for his book on forestry. Under the chapter heading 'What Foresters on Private Estates Are Trying to Do', he wrote: 'Fundamentally it is to make money in a useful industry.'[35]

Tragically, the fate of Dartington's Atlantic oakwoods was not unique. The example set by the estate helped pave the way for a wider assault on Britain's rainforests by the modern forestry industry.

Wilfred Hiley and Leonard Elmhirst sought to influence government forestry policy in their favour and, according to Michael Young, 'kept up an almost continuous attack on the Forestry Commission for not giving more help to private owners'. Leonard's idea for a 'dedication scheme', in which private landowners agreed to dedicate part of their estates for forestry in return for public grants, 'had a strong influence on the Forestry Acts of 1947 and 1951'.[36] In response to post-war subsidies, more and more landowners began converting their 'scrub oaks' and 'unproductive' woods to fast-growing conifers. The Dartington Estate itself was a beneficiary of one of the new 'Scrub Clearing Grants' doled out in the 1950s, which paid for the clearance of 'invading birch, bramble and bracken' in Fingle Woods.[37]

The historian Oliver Rackham estimated that around a third of the country's ancient woodlands were destroyed in the thirty years between 1950 and 1980. The bulk of this loss could be attributed, he found, to modern forestry policies. 'The long arc of destruction has reached from the Lizard Peninsula to remote Argyll,' he wrote, singling out the loss of the 'steep oak woods of mid Wales' and the 'romantic lichen-hung corkscrew oaks of the deep valleys of Dartmoor and Bodmin Moor' as being particularly hard to bear.[38]

It's hard to know precisely how many of Britain's temperate rainforests succumbed to the axe in this period, because Forestry Commission censuses at the time seldom recorded the ecology of a wood, only its suitability for timber production.[39] Thanks to the painstaking work of ecologists and historians like Rackham, however, we now have a set of Ancient Woodland Inventories for England, Wales and Scotland, which allow us to glimpse some of the scale of devastation. The English and Welsh inventories include a category of 'Plantations on Ancient Woodland Sites', or PAWS, recording where very old woods were felled and replanted with timber trees, usually conifers. (The ancient woodland data published by Scottish Natural Heritage doesn't have a separate PAWS category, though clearly many were coniferised in Scotland, too.)[40]

Using digital mapping software, I've measured the area of plantations on ancient woodland in the western oceanic zone.

At least 29,000 acres of ancient woods were felled for plantations in England's oceanic zone, and over 37,000 acres in Wales. We can't be certain, but it's a good bet that many of these ancient woods were temperate rainforests, supporting numerous species of mosses, lichens, ferns, birds and mammals. In other words, we may well have felled some 66,000 acres of Britain's rainforests in the twentieth century alone – an area the size of Birmingham – and all in the name of planting trees for timber. The road to hell, as they say, is paved with good intentions.

On my explorations of Britain's rainforests, I've seen the scars left by modern forestry plantations everywhere. In Cornwall, gorgeous temperate rainforest survives in pockets at Golitha Falls and at Cabilla, on the damp edges of Bodmin Moor. But follow the waters that drain off Bodmin's peat bogs down the River Fowey, and you'll find that most of the ancient woods that formerly clothed this valley have been butchered. From their bones now grow darkly forbidding pines. Many of these plantations are managed by the Forestry Commission, either owned outright by it or leased to it by aristocratic estates hungry for money.

In Wales, where entire valleys were swallowed up by plantations during the twentieth century, many ancient woods also succumbed to colonisation by conifers. Even the green fastness of Y Felenrhyd, the rainforest name-checked in *The Mabinogion*, hasn't escaped the forestry drive unscathed. During our visit there, one of the footpaths took us along the threshold between the centuries-old oak woodland and a modern pine plantation on its edge. The contrast was stark. To our right, old oaks twisted and coiled, their branches playing host to a wealth of epiphytic mosses and lichens. Beneath them grew a rich, scrubby understorey of birch and rowan saplings, bilberry bushes, heather and ferns.

To our left, parallel lines of conifers stood ramrod straight, their bark too acidic to support lichens. The forest floor was largely barren, cloaked in shadow and covered in a thick carpet of fallen pine needles.

This artificial environment – of straight lines, tidy uniformity and trees organised into 'yield classes' to measure their productivity – is a salutary warning of what happens when economists triumph over ecologists. The mindset of the forestry economist, pioneered by Hiley at Dartington, became institutionalised in the Forestry Commission and private forestry firms. As the conservationist Peter Marren observes, 'modern tree-planting has closer affinities with arable agriculture than traditional woodmanship: the ground is ploughed, fertiliser, and sometimes pesticides, is applied . . . and later the crop is harvested in extensive clear-fells'.[41] Traditional methods of woodland management, such as the coppicing practices that allowed some of the old oakwoods around Dartmoor to survive, were jettisoned. Core to the ethos of twentieth-century forestry was control: mankind (and it was usually men, wielding chainsaws and clipboards) bringing order to nature. The veteran environmental campaigner Chris Rose led Friends of the Earth's campaigns in the 1980s against the despoliation of Britain's ancient woodlands. 'I think it has a lot to do with psychology,' he told me. 'In woodmanship, nature was essentially in charge. In UK-style forestry, humans were in charge.'

But, even amidst such destruction, there is cause for hope. Even the tidiest forestry plantation often contains relics of the ancient woodland that lies beneath it. I found this when, on a cold winter's day in late 2021, I went for a trespass with a group of friends in King's Wood. The Dartington Estate no longer owns it, having sold it to a private forestry company some years ago: our approach was met by threatening no access signs. Yet whilst King's Wood remains predominantly a plantation, its regiments of conifers standing stiffly to attention, there were also signs of a mutiny taking place amongst the ranks. Along the edges of the forestry tracks, wood sorrel and bilberries still grew; an oak sapling had taken root. Remnants of the old ecosystem were staging an uprising. Below the conifer crop, the forest floor still contained the buried seed bank of the old rainforest, waiting to burst forth. *Life . . . finds a way.*

Given the role played by Devon's Dartington Estate in pioneering the destruction of our rainforests, it's poetic justice that the county also gave birth to its antithesis. In 1972, a retired Devonshire farmer called Kenneth Watkins founded the Woodland Trust.

Watkins lived and farmed in the valley of the Erme River on southern Dartmoor, just a mile or so south of Piles Copse, one of the three upland oakwoods that survive on the moor. An amateur naturalist, Watkins kept a pet badger called Meles – after the Latin name for the animal – and his favourite place to sit was in a wood on his farm, now recognised as being Atlantic oak woodland, 'abundant with ferns, mosses and lichens'.[42] Watkins was alarmed at the decline of Britain's old woods. So one day, sitting round the kitchen table of his Dartmoor neighbour and fellow naturalist Henry Hurrell, Watkins and a group of friends set up the Woodland Trust, dedicated to protecting ancient woodland and restoring those which had been ravaged by modern forestry.[43]

On 24 October 1972, the newly founded Woodland Trust made its first purchase: a stretch of upland oakwood in the Avon valley, south Devon. With private investors circling, the ancient wood was under threat of being felled and converted to a conifer plantation, so the Trust stepped in to save it from the axe.[44] Further acquisitions in the southwest followed, and before the end of the decade the charity had moved from a regional focus to a nationwide one. Its emergence also struck a chord with the public, with thousands soon becoming paid-up members.

The Trust caught the first wave of the modern environmental movement. The early 1970s witnessed a counter-cultural revolt against damage to the planet which spread across the developed world. In Britain, this also manifested itself as a reaction against the turbo-charged destruction wrought on the countryside since the Second World War by industrialised farming and forestry. It was a movement that united hippies and Tories, young students and old professors. Some of the new environmental activists were inspired by the counsel of an old Oxford don who had seen trees as living beings, rather than as numbers on a balance sheet. It wasn't a coincidence that in the same year as the Woodland Trust was founded, the *Sunday Times* reported the paperback edition of *The Lord of the Rings* to be selling a cool 100,000 copies a year.[45]

Over time, the Woodland Trust sought not just to buy up ancient woods to protect them from coniferisation, but also to restore those that had already succumbed. In 2013, in a delicious twist of history, the Woodland Trust – working jointly with the National Trust – acquired the former Atlantic oakwoods turned into forestry plantations by the Dartington Estate at Fingle Woods.[46] The new owners had their work cut out for them: an early management plan for the woods reported that Hiley's forestry works had 'undoubtedly caused immense damage to their value for biodiversity'.[47] I was keen to learn more about how this felled rainforest might eventually be reassembled, so I arranged to go and visit the site on a wintry November afternoon.

I was met by Dave Rickwood, site manager for Fingle Woods, along with Eleanor Lewis, the site's community engagement officer, and Ross Kennerley, who is the southwest regional manager for the Woodland Trust. After a warming mug of tea in their log-cabin offices – I was shivering from the cold – we began to explore the wood. Our starting point was a block of conifers, its interior inky-black, bisected by a forest track that was letting in some light. Fir trees grow very fast: as Dave explained, 'in the language of foresters, Douglas fir has a "yield class" of 20 to 25, whereas oaks are only about 4.' This means that oaks quickly get swamped by conifers, shading out the sunlight they need. Removing the conifers nearest the surviving oaks gives them space to grow, creates better light conditions for lichens and also allows the ground flora – the bilberries, ferns and heather – to recover.

The first thing to do when restoring an old wood, Dave told me, is to look for linear features – edges, hedges, river banks. These are the places most likely to have escaped the axe, and therefore where the old deciduous trees and understorey plants are most likely to survive. 'Like veins running through the forest, keeping its pulse going,' said Eleanor. The Dartington Estate subdivided Fingle Woods into 179 different compartments, as it wanted to experiment with multiple types of timber trees. To give Hiley some credit, this system has – accidentally – resulted in lots of residual edges where the ancient woodland remnants have clung on.

One of these edges is the River Teign, running through the heart of Fingle Woods. We stood on its banks, admiring the veteran oaks and hazels that line it, shedding their leaves into the river so that it was covered with a floating carpet of copper and bronze. A botanical survey carried out in 2014 found a 'relic assemblage' of rare oceanic lichens to have survived in the woods, with the main areas of interest along the Teign.[48]

As we watched, a brown sea-trout leapt out of the river, startling us with a splash. But the fish in the river are declining, and no one seems quite sure why. Could it be pollution upstream, I wondered aloud, or the acidity of the water? Dave reckoned temperature rises may be to blame, with the climate crisis altering the conditions for when fish can spawn. One of the challenges with restoring an ancient woodland is that the baseline has shifted: the world is already 1°C degree hotter than pre-industrial levels, and could yet rise by 2 or 3°C without further urgent action by governments.

Rising temperatures, and lax controls on the international trade in plants, have also opened a Pandora's box of novel tree diseases. Every year, it seems, a new threat to our trees emerges. In the 1970s, it was Dutch elm disease which wiped out nearly all of Britain's elms; now, the list is far longer, including *Phytophthora ramorum*, a disease affecting larches; the processionary oak moth caterpillar; and ash dieback, which threatens to kill off most of Britain's ash trees. This makes it all the more important to minimise imports of exotic conifer saplings and to grow more native trees from domestic seed stocks.

Our visit proceeded to a part of Fingle Woods which was once planted with larch but which had to be suddenly clear-felled because of a plant health notice from the government warning about the spread of *Phytophthora*. Ordinarily, PAWS restoration is a gradual, laborious process of taking out small areas of conifers and allowing broadleaved trees to take their place, whether through planting or natural regeneration. In this instance, however, the Woodland Trust had been forced to take a more drastic course of action, removing a large area in one fell swoop. The results so far, Dave told me, were encouraging: by ensuring the contractors doing the clear-felling had left behind as many self-seeded native saplings as possible, these were now spreading out to fill the newly available space. Unfortunately, conifers take advantage of the new light conditions to self-seed, too, so the next phase would be weeding these out.

I was amazed to hear that it's possible to *speed up* the ageing process for woods. 'Veteranising' trees can involve knocking off branches to mimic storm damage, creating knot-holes and nooks in which birds nest. At the more extreme end, it can include killing a tree by removing its bark, and injecting the trunk with fungi to create standing dead wood – accelerating the natural processes of death and decay in woods. In forestry plantations, dead trees tend to be quickly weeded out and rotting timber tidied up. But dead wood, as we've learned, is a superb habitat for beetles and other insects, which in turn provide a food source for the insect-eating birds that inhabit our western temperate rainforests, such as the pied flycatcher. A small cluster of oaks in Fingle Woods, Dave points out, supports five or six pairs of this declining species.[49]

Some choices in restoring old woods are not easy to make. The process, by its nature, requires some trees to be felled – mostly the artificially planted conifers, but sometimes also native species that have reached unnaturally high levels. One example is holly, a tree common in British woods but ordinarily kept in check by browsing animals. In plantations undergoing restoration, holly can have a tendency to spread rapidly in the clearings opened up by the removal of conifers and then seed itself across the wood. Being evergreen itself, it too shades out other saplings and understorey plants. On a later visit to another woodland restoration project, I watched with mixed feelings as the team of contractors chainsawed their way through a mass of upstart holly bushes.

As a keen advocate of rewilding, I find myself asking questions about whether we should really be intervening in ecosystems at all. But rewilding isn't simply about 'letting go': that would mean walking away from the mess left by previous generations. For rewilding to work, it still requires people to step in at the start of the restoration process, to try to correct the errors humans made – such as by taking out plantation conifers – and set in train natural processes so that the ecosystem can take care of itself again. We can recruit other animals to help with this, and reintroduce missing species, as explored in the previous chapter. But we may also find that there are some ways that people have historically interacted with woods – such as traditional coppicing practices – that are worth reviving, alongside truly 'wild' woodlands.

'So how long could it take to restore Fingle Woods to the

temperate rainforest it once was?' I asked. Dave paused, and thought for a moment. 'In terms of getting to a stage where all the conifers have been removed . . . it could be eighty to a hundred years,' he said, a little wearily. And for the lichens to start reappearing, particularly the Lobarion species of old woods, it could take even longer. Cutting down a rainforest, sadly, is much quicker than putting one back.

Fingle Woods is not the only felled rainforest being slowly pieced back together. In 2020, the Woodland Trust and National Trust joined forces again to buy Ausewell Wood on the Dart, a site they described in their fundraising appeal as being 'like a lost world', specifically referring to it as 'temperate rainforest'.[50] It's a truly wonderful place, and I've spent many hours exploring its craggy oaks and seeking out the stinky *Sticta* lichens that grow on some of them. But the enjoyment is matched by a sense of horror at knowing what must have been lost on this spot: the majority of Ausewell is a drab plantation. Once, Louisa and I decided to make the long, steep descent from Ausewell Rocks down to the River Dart far below. The contrast between the green, airy fragments of oakwood, filled with birdsong, and the gloomy, lifeless understorey of the pines was profoundly depressing. But, under its new owners, Ausewell now has a chance to be restored to its former glory.

Institutional mindsets have changed, too. The Forestry Commission had a change of heart in the 1980s, introducing new grants focused on planting broadleaved trees rather than conifers, and nowadays it has a policy of restoring Plantations on Ancient Woodland Sites that are within its ownership.[51] Private plantation owners, however, are under no such obligation. There are around 92,000 hectares of privately owned Plantations on Ancient Woodland Sites in England.[52] Yet in 2020, private landowners couldn't be bothered to restore more than 67 hectares of them.[53]

Some owners of felled rainforests remain very resistant to

restoring them, too. Just to the south of King's Wood and Ausewell lies Dean Wood, yet another former temperate rainforest felled for forestry in the twentieth century. Its current owner, Robert White, a former director of commercial forestry firm Fountains Forestry, has no intention of restoring his plantation to rainforest. 'I'm not in favour of the wholesale felling of high quality conifer on old hardwood sites and their replacement with low grade hardwood,' he stated in 2010.[54] To some foresters still, 'Atlantic rainforest' is merely 'low grade hardwood'.

Very occasionally, however, former rainforests come up for sale, holding out the prospect that they can be bought and restored. And during the writing of this book, I was fortunate enough to be in exactly the right place at exactly the right time to help out with one such effort.

One mild October evening in 2021, whilst visiting my parents in Cornwall, I went for a walk. The sky was lit up with a spectacular sunset, colouring the white clouds with shades of peach and lavender, as I headed down the hill towards High Wood. My grandparents had farmed in this area since the Second World War, and my Gran – whom we all knew as Nanny – used to go for walks in High Wood when it had been an ancient oak woodland. Nanny loved nature, greeted birds as old friends and was a true devotee of David Attenborough documentaries; once, not long after she'd turned ninety-three, I found her putting up an Extinction Rebellion poster in her window. Polypody ferns grow luxuriantly on trees in this part of Cornwall, and I once discovered some tree lungwort growing on an old oak just to the north; so High Wood may well once have been temperate rainforest. But in the 1950s and 60s, it was coniferised by its owner at the time, the Duchy of Cornwall, and became a forestry plantation.

When I reached the entrance to High Wood, it was almost dark, but there was enough light to make out a sign on the gate that hadn't been there on my previous visits: WOODLAND FOR SALE. My heart skipped a beat. I'd been hoping for years that High Wood might come on to the market, so that someone could at last take on the job of restoring it. As soon as I got home, I looked up the sales particulars for the wood and hurriedly sent a tweet: 'Wanted: a buyer to restore a lost rainforest to its former glory.' There was just one catch: the seller was asking for £400,000 by the end of the week.[55]

The next morning, I woke up to a string of Twitter notifications, most of them saying things like 'Have you considered crowdfunding?' and 'I'd love to help – if only I had the money.' With the sales deadline just days away, there was never going to be time to raise the funds from people chipping in a fiver at a time. One tweet, however, was different. 'Happy to buy and donate,' it read simply. At first, I almost scrolled past it. But then I googled who it was from, my eyes widening as the results came in. *Sunday Times Rich List . . . Investor profile . . . Net worth . . .* My tweet had been seen by a multimillionaire financier, who seemed to be genuinely offering to stump up the requisite cash. Half thinking I was still dreaming, I slid casually into his DMs.

There followed a frenetic forty-eight hours of back-and-forth messages, phone calls and emails, as I tried to help connect this incredibly generous donor – who wished to remain anonymous – with an organisation willing to take on the restoration of High Wood. I called up Ross Kennerley at the Woodland Trust, who scrambled together a meeting of their property department at zero notice to consider the options. But there's a limit even to what the Trust can do, and saddled with the running costs of managing over a thousand woodlands nowadays, it can't take on every new site that comes up. After an apologetic phone call from Ross, I was back to the drawing board.

Then, salvation. I was contacted by an enthusiastic young tree-planter, Phil Sturgeon, whose tiny charity, Protect Earth, was prepared to take on managing High Wood. With the clock ticking down towards the deadline for bids, I connected him to the anonymous donor, jumped on a Zoom call to pass on as much information as I had about the wood's condition, and crossed my fingers as they submitted their offer to the estate agents. Days passed. And then a call from Phil, excitedly telling me their offer had been accepted. I breathed out a long sigh of relief. High Wood had been saved – or rather, its fate was no longer sealed. It had been handed a fresh chance. The real work of restoration, the hard slog of painstakingly putting it back together, would take decades.

About a month later, I went to meet Phil for a walk around High Wood, bringing along Mum and Dad so they could relate some of the history of the site. Nanny had been unwell and in hospital for some weeks, and Mum was due to visit her that afternoon. As we walked down the footpath through High Wood, Mum got the awful phone call saying that Nanny had passed away. As we hugged one another, tears coursing down our cheeks, there was a keening cry overhead. A beautiful buzzard was circling the wood, catching the rising thermals with joyful abandon. It seemed in that instant to embody my gran's spirit, soaring free, reunited with nature. I hoped that, from where she was looking down, she would approve of the work now being done to restore a fallen rainforest.

8

The Accidental Rainforest

The road to one of England's most spectacular lost valleys was narrow and winding. November rains cascaded down lanes and formed puddles in neighbouring fields. Our car careened past steep-banked hedges, bouncing over crater-sized potholes, struggling with the frequent changes of gear demanded by Devon's hills. There are few passing places on these roads, and it pays to be fast at reversing, particularly when an unforgiving farmer in their tractor is bearing down upon you. Hedgerow ferns and brambles clipped our wing mirrors and flailed through our open windows, trying to cadge a lift.

We veered sharply to the right, then plunged down into what felt like a green tunnel, Louisa and I leaning into the windscreen to gaze up at a ceiling formed from the canopies of trees. It reminded me of a picture book I read as a kid: a modern fairy-tale in which the central character takes a wrong turning off a motorway's spaghetti junction and emerges in a magical land. Up ahead, a signpost loomed, almost lost in the undergrowth, pointing the way to our destination: LUSTLEIGH.

It was our friend Sam Lee, a folk singer and environmentalist, who had first told us about the place that would come to fascinate me. One day, not long after we'd moved to Devon, Sam visited and took Louisa and I wild swimming at a spot he knew on the Dart. Later, drying out in the sun on the banks of the river, we pored over maps of Dartmoor, whilst Sam pointed out his recommendations for places to see. 'You *have* to go there,' he said, his finger hovering over an area marked in green, a wood surrounded by the steep contour lines of a ravine.

I remember taking the map for a closer peek, thinking how unusual it looked. Many of the large forests in Devon nowadays are conifer plantations, but this was a great expanse of deciduous woodland, spreading along what Devonians call a *cleave* – a deep, steep-sided gorge. *Lustleigh Cleave*. I rolled the words over in my mouth, and those that it brought to mind: *lustrous, lusty, luscious*. It sounded as mysterious as it was appealing. Remote from any main roads, accessible only by footpaths, it looked like the land that time forgot. Sam was right: I had to go there.

We parked at one of the tiny hamlets that cluster on the outskirts of Lustleigh Cleave. This one had the unlikely name of Water, and we soon found out why. Water defines the Cleave, shaping its geology and its climate. The River Bovey carves its way through the valley, fed by several streams that wend their way down its sides, suffusing the air with a cool humidity. After a rainstorm, these waterways are joined by numerous unofficial tributaries which course along the many green holloways leading into the Cleave. It was down one of these, a footpath that had become a torrent, that we made our first venture into Lustleigh Cleave.

The high banks of this holloway abounded in plants: the crisp, fleshy leaves of pennywort; the clover-like trefoil of wood sorrel; a tangle of bramble and ivy, bracken and holly, studded with the verdant sheen of hart's-tongue fern. Crumbling stone walls were braided with the splendidly named fern maidenhair spleenwort, its fronds like emerald necklaces. Great oaks and ashes lined the sunken lane, their fern-covered branches forming a green roof: like couples at a ceilidh making arches with their arms as they dance Strip the Willow. Some of the oaks had vast buttress roots, making me think of photos I'd seen of rainforest trees in the Amazon. Moss overlaid every surface like a deep-pile carpet.

As we walked, the holloway led us deeper and deeper into the Cleave, down into the bowels of the earth. And the further we descended, the damper it felt. I began to notice species appearing that marked these woods out as rainforest: flowery masses of *Peltigera*, or dog lichen, and the musty smell of *Sticta*. A tangled blue-green mass lying on the forest floor turned out to be a beard lichen, one of the *Usnea* species, which had fallen from the canopy. High up in the arms of a beech tree, a tiny sapling sprouted, its aerial roots nourished by rainwater. A gigantic holly grew octopus-like from an embankment, its tentacular limbs – each thicker than a person's torso – resembling an awakening Kraken.

The woods grew denser and darker around us, a crisscrossing latticework of green trunks and branches, glimmering in the gentle drizzle. But through the trees, at points, we could glimpse the high horizon: though we were still travelling down towards the valley floor, we could see ahead of us the other side of the Cleave rising up like a cresting wave. I was struck by how big these woods were: the far better-known Wistman's Wood is a mere eight acres, but those at Lustleigh span hundreds. Yet hardly anyone I've asked has ever heard of them.

Picking our way over rocks, Louisa and I reached the bottom of the Cleave, crossed the River Bovey via a narrow footbridge, and began to ascend the other side. We soon noticed the Cleave's slopes here were covered in moss-encrusted boulders, some of them as big as cars. Further upstream, we'd later find, some of these boulders had fallen to form a dense carapace over the Bovey itself – making the river possible to ford by simply stepping from rock to rock, the currents cascading beneath them.

But as we walked further, we came to a jumble of stones

which didn't look natural. Venturing closer, it became obvious that beneath the mosses and ferns lay tumbledown walls. Hazels and birch trees sprouted from cracks between the slabs. We were standing in the ruins of a farmstead now overgrown by a forest. It was a sight I associated more with images of Mayan temples in Latin America as they succumbed to the smothering jungle. The surrounding farmland had clearly been abandoned and was being reclaimed by regenerating woods. I began to realise that we'd stumbled across something extraordinary: an accidental rainforest.

Over the following months, I would return to Lustleigh Cleave again and again, captivated by its beauty and its mysterious past. It kept pulling me back, tantalising me with its secret ruins in the woods. My visits and research led me to eventually arrange a meeting with the one person who I knew held the key to unlocking this mystery: the village archivist.

The village of Lustleigh itself (population: 600) is achingly picturesque, the sort of place you'd set an episode of *Midsomer Murders*. The cottages are thatched, their walls whitewashed; the village green is well manicured, overlooked by tearooms and the local pub. A local history pamphlet for sale in Lustleigh's 800-year-old parish church proudly records that, at the time of the Domesday Book, the settlement was called *Sutreworde* and possessed 'a very large area of forest'.[1] But despite this long association with woods, the quaint tidiness of Lustleigh village seems a world away from the wildness of the Cleave, which lurks out of sight, on the other side of the hill.

Louisa and I arrived in Lustleigh one sunny spring day to meet Peter Mason, local historian and long-time member of the local history society. The tiny churchyard was neatly mown, yet I couldn't help noticing that many of the names on the gravestones were obscured by a thick mat of lichens. The society's archives are stored in the Old Vestry, and there we met

177

Peter: an avuncular historian with grey hair, wearing spectacles and a fleece jacket. I'd later return to the vestry to pore over old documents; but first, Peter wanted to show us a different view of the Cleave – one that would give us a glimpse back through time.

After a fifteen-minute walk through the village and up the hill, we arrived at a deep holloway, filled with uncoiling fern croziers and studded with the coin-shaped leaves of pennywort. Bluebells shimmered in a sunlit glade. The wooden footpath-sign pointing us to the Cleave had succumbed to at least three different species of lichen, creating a mosaic of yellows, greens and greys. I stopped to watch a shiny black dung beetle amble across the path, its wing casings iridescent, flashing blue and purple as it moved. High above, the oak canopy glowed like stained glass, the leaves a startling shade of lime-green.

Vast granite boulders loomed out of the woods, covered in ivy and moss. We were climbing up a tor: one of the wind-worn granite outcrops that are found across Dartmoor, protruding through its covering of peat and moor-grasses. Unusually for Dartmoor, however, this tor was almost completely covered in trees. At its summit lay a collection of rock pillars, largely obscured from view by the burgeoning wood. A clearing around a pile of rounded boulders was illuminated by sunlight, and from here we gazed down upon Lustleigh Cleave, its spreading expanse of temperate rainforest filling our field of vision. On the other side of the combe lay the ancient woods that Louisa and I had walked through on our first visit to the Cleave. But the oaks and hazels on this side of the river, washing up towards the summit like a tide, were much more recent.

Peter rummaged in his rucksack, and brought out a folder of laminated images. 'This is what the Cleave *used* to look like,' he said, showing us the first image. It was a printout of an oil painting by the Devon artist Francis Stevens, completed in 1820. Entitled simply *Lustleigh Cleave*, Stevens's artwork depicted a landscape utterly unlike the one in front of us. The Cleave of two centuries ago appeared as a sweep of bare hillsides, capped by a rocky tor bereft of all vegetation. On the horizon, Stevens had painted in a thin plume of ascending smoke, a sign of the old Dartmoor custom of swaling. In the painting's foreground, a shepherd lounged on one of the boulders atop the Cleave, his flock of sheep grazing at his feet. One thing hadn't changed, however: Stevens had captured well Dartmoor's inclement weather, showing sheets of rain descending from lowering clouds.[2]

'Let's spool forward a century,' said Peter, and brought out a second image, this time an Edwardian postcard from a time when Lustleigh Cleave had become a huge tourist attraction. The railway branch line that once brought sightseers to Lustleigh from the bright lights of Newton Abbot fell long ago to Beeching's axe, which resulted in the closure of thousands of miles of track. But in the 1900s, the Cleave drew such numbers of visitors that it generated a veritable cottage industry of picture postcards, capturing it from all angles. Preserved in these early colour photos, tinted yet faded, is the story of a landscape regenerating. The one Peter held was taken from the same standpoint as the canvas painted a hundred years earlier, but now vegetation was rising up around the tor. Bushes had colonised the upper slopes and bracken was starting to engulf the boulders. Two Edwardian women in their Sunday best occupied the space once grazed by the shepherd's flock.

'Now look where we are,' Peter said, lowering the photo. We stared at the dense forest before us. I looked back quizzically at Peter. 'See the outlines of that tor?' he asked, pointing at a distinctively shaped pillar of rock present in both the painting and the postcard. 'It's over there.' We craned our necks to peer through the tangle of trunks and branches. There, just faintly visible amidst the undergrowth, was the rock, our anchor-point to the past.

'That's amazing!' I laughed. 'Can I borrow those?' Taking the two pictures, I held them in front of me, trying to line up the outlines of the rocks, flicking back and forth through time. Peter and Louisa watched bemused as I darted through the undergrowth towards the granite pillar, and then back again, trying to find the exact spot where the shepherd had formerly sat guarding his sheep. But there were too many trees in the way. Where sheep had once grazed upon a wet desert, 300 acres of lush rainforest now grew.

A mountain of other evidence highlights the profound transformation of Lustleigh Cleave over the past two centuries. At first, the changes were only gradual. When the antiquarian vicar Samuel Rowe visited the Cleave on his perambulation of Dartmoor in 1827, he described it as a 'rocky labyrinth', comprising 'picturesque masses of rock with shrubs and foliage springing up from their fissures'.[3] The first and second edition Ordnance Survey maps, both published in the nineteenth century, show Lustleigh Cleave as being barren of trees.[4] And when the writer William Crossing featured the Cleave in his 1909 guide, he was still able to describe it as being 'bare'.[5]

But then the process of natural regeneration seems to have picked up pace. The RAF's aerial photo survey of Britain shortly after the Second World War gives us the first images of Lustleigh Cleave from the air, revealing a clearly changing landscape.[6] Trees and scrub have made the jump across the Bovey River and started marching inexorably up the hillside. Another fly-by photo taken in the early 1960s shows the encroaching vegetation gathering pace;[7] whilst a tourist guide to Dartmoor a decade later describes the Cleave as being 'covered with bracken, heather and scrub', though noting that 'the tree-growth peters out on higher ground'.[8]

In the past fifty years, Lustleigh Cleave has undergone a sea change, turning into something rich and strange. A soil scientist who first visited the area in the mid-1960s remembers it then being 'open, almost savannah-like grassland', but by the start of the twenty-first century, 'young oak-birch woodland had developed, obscuring the rocks and the view over any distance'.[9] Peter Mason, too, has borne witness to these changes during his lifetime. Pictures he took of Lustleigh Cleave from when he first moved to the area in the 1970s reveal a much more open landscape, grazed by Dartmoor ponies. A glance at the aerial photos available on Google Earth show the rainforest's advance has accelerated even in the past two decades.[10]

And what of the farmhouse buried in the forest? On old maps, it bears the name of Boveycombe; and local archives suggest the last person to inhabit it was one George Crocker, who grew potatoes here in the 1940s for the war effort. But nature has since reclaimed these fields dug for victory. In 2011, the ruins of Boveycombe farm were unearthed by a team of archaeologists, who peeled back the mat of mosses covering its stones to peer into a long-gone past.[11]

Piecing together this process of transformation fascinated me, because it spoke to how often we fail to notice the profound changes taking place in nature around us. Environmentalists talk about 'shifting baselines' to describe how people get used to ecological decline: we become so accustomed to the waning of species that we can forget even what we've lost. But it works the other way, too. We can also overlook the gradual resurgence of nature, and forget its powers of renewal. For whilst the world had turned its back on Lustleigh, a rainforest had returned to this corner of England.

Why, then, has rainforest been able to regenerate on Lustleigh Cleave? I became captivated by this question. Finding an answer to it matters – not just to understand what's happened in this specific part of Devon, but because it could have huge ramifications for other places in Britain like it. Had the rainforest returned due to deliberate human effort, I wondered? Or had the area been abandoned, due to some economic disaster or collapse in population?

It turns out that Lustleigh Cleave *did* experience a population crash in the 1950s – except the species that collapsed wasn't humans, but rabbits. Rabbits were first introduced to Britain by the Romans, then bred for fur and meat in medieval times.[12] As rabbit numbers increased, so did their ecological impacts: rabbits love to nibble young saplings, preventing the natural regeneration of trees. And because rabbits breed like, well, *rabbits*, their populations reached plague-like proportions by the early twentieth century. Then myxomatosis struck.

Myxomatosis is a horrible disease that causes rabbits to go blind before they die, their bodies covered in swellings. In *Watership Down*, the rabbit characters call it the 'white blindness', referring to the pearly sheen of an eye succumbing to the *Myxoma* virus. It was first introduced deliberately in Australia, to control the exploding numbers of rabbits brought over by European colonisers. A private landowner released it in France in 1952, and the following autumn, Queen Elizabeth's coronation year, it hopped the Channel to Britain. Myxomatosis ripped through the British countryside like wildfire, causing the rabbit population to crash by 95 per cent. Winston Churchill, then Prime Minister, tried to contain it by making it a criminal offence to intentionally release the virus, but it was to little avail.

Horrifying though this disease was, one consequence of the spread of myxomatosis was a rebound in the vegetation that rabbits ate. Fields previously nibbled to the quick by the once-burgeoning rabbit population saw their grassy swards grow taller, with white clover and other wildflowers staging a recovery. 'The decimation of the rabbit had profound implications on the balance of nature,' concludes one recent study.[13] The disease may also have kickstarted the resurgence of rainforest on Lustleigh Cleave. A parish newsletter clipping I found in the

Lustleigh Society archives from 1985 made the case for this. 'Before the advent of myxomatosis, which wiped out the rabbit population on the Cleave, there were hardly any scrub oaks and birches and few serious brambles,' the article claimed. 'Once rabbits died out, seedling trees grew unmolested'.[14]

But the more I read, the less I became convinced that this was the real answer. Though rabbits are formidable nibblers of tree saplings, sheep and cattle are still more so. And Lustleigh Cleave has been grazed by sheep for hundreds of years – as we saw in the early nineteenth-century painting, with its flock of sheep grazing on the bare summit of the Cleave. Instead, the fundamental reason why this landscape has transformed from barren hillside to lush rainforest lies in the changing intensity of livestock grazing. Which leads us into the heart of the matter, a set of interlocking conflicts fraught with politics. Because Lustleigh Cleave is a common.

A common is an area of land which is often privately owned, but over which various other people – known as commoners – have usage rights. 'Common rights' cover a whole host of ways that commoners can use common land in order to draw sustenance from it. They can include the right to fish in a river (which, in medieval times, was called *piscary*); the right to cut peat for fuel (or *turbary*), to collect firewood (*estovers*), and to put pigs in a wood to eat acorns (*pannage*); and, most importantly of all, the right to graze livestock on common land. Rights of common are an ancient set of customs, dating back to before the Norman Conquest. Common land covered about 30 per cent of England at the time of Shakespeare, but the process of 'enclosure' – the privatisation of the commons by the landed gentry – led to their decimation: only about 3 per cent of England nowadays is common land.[15]

Commons might therefore seem like something of an irrelevant anachronism in modern Britain. But 3 per cent of England is still roughly a million acres, and a lot of that lies within the western upland areas that enjoy a rainforest climate. Commons also cover nearly 300,000 acres of Wales, and 1.3 million acres of Scotland, with particular concentrations in the rainy Welsh uplands and the hyper-oceanic Highlands and Islands.[16] So getting to grips with how these complex spaces work is important if we want to restore Britain's rainforests.

What's more, debates about how common land is best governed lie at the heart of modern environmentalism. The debate started in the late 1960s, when the ecologist Garrett Hardin published a now-infamous paper called 'The Tragedy of the Commons'. Hardin argued that individuals are always compelled by self-interest to overuse 'common pool resources' like common grazing land. 'Each man [sic] is locked into a system that compels him to increase his herd without limit – in a world that is limited,' Hardin declared. The tragedy of the commons was that common land would inevitably end up exhausted and depleted by individual commoners each drawing more than their fair share. Hardin saw only two ways to prevent overuse: coercive intervention by the state, or the privatisation of common land so that individual property owners were incentivised to practice self-restraint. As a liberal who prized individual freedoms, Hardin favoured privatisation – essentially the final enclosure of the remaining commons.[17]

Hardin's diagnosis has rightly drawn much scorn over the decades, not least from the Nobel Prize-winning economist Elinor Ostrom. Drawing on studies from around the world, she pointed out that many commons had been sustainably managed throughout history via local systems of communal governance, without the need for either state control or private property.[18] In Britain, some of these ancient intermediary institutions for managing commons still exist, like the Verderers of the New Forest.[19]

Most discussions about commons end here, left in the realms of history and theory. People's views on commons, if they have any view on them at all, tend to split down ideological lines. Left-wing economists and environmentalists like Guy Standing, George Monbiot and Peter Linebaugh have championed the commons as an alternative to both the free market and the state, arguing that commoning breeds 'mutuality and negotiation', 'destroys inequality' and 'provides an incentive to protect the living world'.[20] As a green lefty, I've always thought that the enclosure of common land was a huge historic injustice, and believed in the ideal of commons as a way of managing shared resources both fairly and sustainably. I still think so. But to really understand the intricacies of commons, it's essential to look at how they actually work in practice in modern Britain.

The truth is that many of the overgrazed uplands of Britain – those which George Monbiot pithily describes as being 'sheep-wrecked' – are also commons. The overgrazing they suffer from can be traced directly to a piece of legislation introduced by the UK government which – ironically – set out to better regulate commons: the 1965 Commons Registration Act. The formal registration of common rights had been recommended by a Royal Commission as a tidying-up exercise; but rather than improve matters, it significantly worsened them. The Act formalised the number of livestock each commoner could graze as legal property rights, generating a bidding war amongst commoners. Where once local communities had policed the number of livestock on each common, now the state sought to step in by getting everyone to write down how many rights of common they held. Suddenly, every commoner felt obliged to register as many grazing rights as they could, lest they lose out to their neighbours. Landlords also encouraged their tenant farmers to claim more rights, seeing that it would result in more rental income for them. In seeking to regulate the commons, the British government had triggered a free-for-all.

We can see the results by looking at the vast number of common grazing rights registered for Lustleigh Cleave. Around thirty-five commoners between them hold rights to graze 228 ponies, 270 cows and no fewer than 1,451 sheep on the common's 300 acres: almost 2,000 animals in total, or six for every acre of land.[21] By all accounts, grazing this many livestock on that much poor-quality grassland would have strained the carrying capacity of Lustleigh common. One study suggests that the sustainable stocking density of sheep on good-quality upland grass is just two sheep per acre per year, and even that level is too high for heather and other vegetation to regenerate.[22] Had Lustleigh's commoners ever exercised all the grazing rights they registered under the 1965 Act, the Cleave today would be a wet desert, rather than a beautiful rainforest. The modern governance of commons has unsustainability baked in.

Nor are modern commons exactly paragons of equality and fairness. To some, nowadays, the word 'commoner' may bring to mind Monty Python's anarcho-syndicalist peasants: communes of the rural working class, eking out a subsistence from the mud. For much of their history, commons were undoubtedly a form of medieval welfare state, allowing the poorest in society to benefit from the safety net provided by the local common. Yet the reality is now somewhat different. On Dartmoor, rights of common are attached to a property – such as a farmstead – meaning that anyone who buys a property bearing those rights becomes a commoner, with rights to graze on common land nearby. In other words, possession of common rights rests with a relatively small number of property owners, and comes with a hefty price-tag tied to spiralling house prices. Over the decades, as pretty Devon villages like Lustleigh have become magnets for retirees and wealthy second-home owners, the social status of their commoners has gone decidedly upper class.

So who were the commoners of Lustleigh Cleave at the time of its transformation into rainforest? And what did they think of the changes they were witnessing? Leafing through the archives of the Lustleigh Commoners Association, much of it typed in faded ink on musty sheets of paper, I started to get a sense of some of the colourful characters who were commoners here in the 1980s and 90s, many of them now long gone. What emerged from these documents was the story of a quintessentially English group of retired colonels and local farmers, goat-herders and members of the landed gentry – united in defending their ancient rights against what they saw as meddling officialdom on the one hand, and the encroachment of nature on the other.

As an ensemble cast, Lustleigh's commoners might have emerged from the pages of an Agatha Christie novel.

They included Lieutenant-Colonel Hugo Pellew, owner of a nearby farm, who once served as aide-de-camp to the Governor of Nigeria during the waning years of the British Empire.[23] He was one of *two* retired colonels in the Commoners Association, alongside a retired captain who had served in the Royal Marines.[24] Some of Lustleigh's commoners were drawn directly from the ranks of the aristocracy, such as Lady Susan Connell, daughter of an earl, who held rights to graze four cows and fifteen sheep on the Cleave.[25] By the close of the century, the landowning establishment had been joined by an increasing number of well-to-do incomers to the area who, having purchased nice, thatched cottages, suddenly found themselves owners of common rights, too. One such unlikely commoner was the brother of *Newsnight* anchor Jeremy Paxman, who moved to Lustleigh and acquired the rights to graze thirty-one sheep and five ponies on the common.[26] Full-time farmers brushed shoulders with hobbyists, old-age pensioners and couples looking to live the Good Life.

The aims of the Lustleigh Commoners Association during this period were simple: its stated objectives were, firstly, 'to maintain our grazing rights', and, secondly, 'to manage the encroachment of gorse, brambles, bracken . . . and saplings' on the Cleave.[27] Some of Lustleigh's commoners appeared ready to do battle in defence of these rights. 'It is an indisputable fact,' declared Colonel Pellew, presumably turning a shade of puce at the time, 'that certain persons . . . claimed Common Rights of grazing, fishing, turbary and estovers, on Lustleigh Cleave under the Commons Registration Act of 1965 . . . Until Parliament repeals or amends this Act, it will remain in force.'[28] For the aged colonel, this appeared to be more a matter of principle than of practical interest: he held rights to graze just a single cow, three sheep and a pony on the common.

But the commoners became frustrated, because their activities seemed to be increasingly hemmed in by bureaucratic red tape. In 1951, Lustleigh Cleave became part of the new Dartmoor National Park, and it was designated a Site of Special Scientific Interest (or SSSI) in 1963. Neither designation prevented grazing on the common, but they did generate a growing welter of management plans, alphabet-soup acronyms, and visiting officials keen to monitor the site. Some of the commoners began railing against what they saw as pettifogging officialdom. Jill Salmon, who kept goats on the Cleave and later became the clerk of the Commoners Association, lambasted the 'administrative arrogance of the bureaucrats who run the Dartmoor National Park'.[29] Colonel Pellew, according to one set of meeting minutes, 'criticised the autocratic attitude of English Nature' – the government's nature watchdog at the time – 'and urged that the Association stand up to them'.[30] There doesn't seem to have been a great deal of sympathy amongst the commoners for what these bodies were trying to do; as one meeting note complained, apparently without irony: 'English Nature seemed to be only interested in wildlife and heather.'[31] Reading the records, their protests struck me as having more in common with UKIP than the Peasants' Revolt.

But in reality, the conservation bodies weren't the enemy. In the early 1980s, a government ecologist wrote to the Commoners Association to reassure them that 'our interest [is] in keeping the balance between trees and open ground'.[32] Some level of continued grazing would, after all, benefit rare woodland-edge species like the high brown fritillary butterfly. Still, there was one stipulation. Under the recently passed Wildlife and Countryside Act of 1981, the commoners would need to create a new management plan for Lustleigh Cleave. In order to do so, they'd have to get agreement with the owner of the common.

The trouble was, no one knew who actually owned it.

The ensuing dispute over who owned Lustleigh Cleave dragged on for years, becoming increasingly arcane. No one, it turned out, had officially claimed ownership of the common via the government's Land Registry. And because the process of registering common rights in the 1960s had been such a mess, with commoners and landowners lodging all sorts of false and conflicting claims, the government ended up appointing quasi-judicial Commons Commissioners to arbitrate between warring parties. In 1982, a Commissioner arrived to hold court on the matter of Lustleigh, hearing representations from three different firms of solicitors, a genealogist from *Debrett's Peerage* (the guide to the landed classes) and, of course, the redoubtable Colonel Pellew. What emerged was a byzantine ownership arrangement dating back some 400 years, in which one-twelfth of the common belonged to the owners of a nearby farm, three-twelfths was owned by another local resident, four-twelfths had been inherited by the Earls of Ilchester, and the remaining one-third claimed by a man who died during the course of the proceedings. Throwing up his hands in horror at this tangled web, the Commons Commissioner eventually ruled that Lustleigh Cleave's legal ownership be vested in an official body called the Public Trustee, with the other claimants becoming beneficial owners. Only in England could you imagine such an anachronistic carry-on.[33]

The delays caused by this debacle, however, had real-world consequences. Peter Mason recalls the ownership dispute triggering the removal of livestock from Lustleigh Cleave, as commoners waited to hear the outcome. Even the Commissioner's ruling brought no immediate resolution, because the Public Trustee found itself unable to agree to a new management plan for the common without the consent of all of the beneficial owners. The Commoners Association became dormant, its meetings tailing off before halting abruptly in 1985. But time and tide wait for no man, and nor does a spreading rainforest.

By the time the commoners eventually reconvened some eight years later, bracken and gorse had crested the summit of the Cleave, with advancing saplings following swiftly behind. There was still no management plan in place. But a few members of the Association appeared determined to take a final stand against the encroaching jungle. Work parties were organised to clear the undergrowth, 'annual bracken crushing' was instigated for the open grazing areas and the commoners reinstated swaling to set fire to the gorse.[34] Slash-and-burn sessions were accompanied by tea and sandwiches. Plumes of smoke rose over the Cleave again, just as they had a century and a half previously.

This time around, however, the commoners were fighting a losing battle. Seemingly in desperation, they consulted one Professor Roy Brown, of the portentously named Bracken Advisory Commission, on how to defeat the insurgent jungle. The professor took the high-octane step of chartering a helicopter to fly a reconnaissance mission over the Cleave and observe the forces arrayed against the commoners. His resulting report described in bellicose terms the 'invading tree areas' and the 'increasing aggressiveness of the bracken', and recommended a 'burning / cutting / ground based chemical control programme' to bring the rebellious vegetation to heel.[35]

At the time, something of a moral panic had been whipped up around bracken. Stories in the Westcountry press warned of bracken's advance across Britain's uplands, calling it 'one of the most dangerous plants on earth', based on some decidedly shaky evidence about the health impacts of its spores.[36] The remedy Professor Brown prescribed was 'aerial spraying with a chemical compound' called Asulam, which Lustleigh's commoners deliberated about using on the Cleave.[37] Colonels and helicopters, invasions and defoliants: I sometimes wondered whether the commoners of Lustleigh Cleave were trying to refight the Vietnam War. *I love the smell of Asulam in the morning*. But in this war of attrition against nature, there was only ever likely to be one eventual winner.

In fact, the cause of keeping Lustleigh Cleave clear of trees ultimately failed for a very simple reason: too few commoners actually wanted to graze it any more. Throughout the 1980s and 90s, hardly any of the local residents who owned common grazing rights chose to exercise them; those who did were a vocal minority. 'I only know of one commoner who is seriously interested in grazing the Cleave,' wrote a government ecologist to the clerk of the commoners back in 1981. 'Any attempt to control bracken or scrub without subsequent grazing will, as you say, fail in the long run'.[38] Over a decade later, the situation remained the same; as a letter from the Commoners Association to English Nature admitted, 'only one commoner is at present exercising grazing rights'.[39] Because rights of common belonged to whoever bought properties around Lustleigh, the commoners were increasingly people who'd moved to different areas to retire or who had other jobs. They had little interest in suddenly taking up livestock herding, and the few active graziers were unlikely to persuade them otherwise. Fast forward to 2022, and today there is still only one active commoner left on Lustleigh Cleave.

But there's also a happier ending for the characters in this story. Despite the reduction in grazing on the Cleave in recent decades, the commoners have still been able to draw an income from it. That's been due to shifts in public farm subsidies over the past thirty years. Up until the late 1980s, the system of

European farm payments – called the Common Agricultural Policy, or CAP – was focused almost entirely on subsidising food production, with payments based on the number of live-stock you owned or the area of land you farmed. But with a growing recognition of the damage being done to the country-side by intensive farming – not to mention overproduction leading to so-called 'milk lakes' and 'butter mountains' – the EU started to switch tack, and began to pay farmers to protect the environment. At the same time that things were coming to a head on Lustleigh Cleave during the mid-1990s, the common was entered into one of the first 'green' farm payment schemes, the Environmentally Sensitive Areas programme.[40] Later, in 2008, the Lustleigh Commoners Association received a ten-year Environmental Stewardship agreement worth £216,000.[41]

For decades, commoners had sought to defend their grazing rights, and the income this brought in, by fighting back the encroaching rainforest. Now, at long last, they were being paid to keep grazing levels low. The war was over.

What does the accidental rainforest of Lustleigh Cleave tell us? It may seem an eccentric story, even a parochial one. But it also contains some profound lessons for rainforest restoration across Britain.

Most importantly, it demonstrates how quickly woodland can naturally regenerate when it's given half a chance. As Oliver Rackham once remarked, 'In England, trees grow where people have not prevented them.'[42] Grazing and swaling prevented trees from growing on Lustleigh Cleave for centuries; but when they ceased, trees sprang back. The ancient woods on the western side of the Bovey seeded saplings that forded the river and spread rapidly up the Cleave. Though the process of natural regeneration here may have started over a century ago, the bulk of it seems to have taken place in the last thirty or forty years. And where the trees have spread, the lichens and bryophytes are now following. Oaks on the summit of the Cleave are now garlanded with string-of-sausages lichens, and I've found *Sticta* colonising hazel clumps on the western banks of the Bovey.

Whilst Lustleigh is perhaps unique in terms of the scale of natural regeneration taking place, it's also far from being the only common where trees are returning as livestock numbers have declined. I've visited various commons around the edges of Dartmoor and Bodmin Moor which are 'scrubbing up' with hawthorn, rowan and willow. The writer Matthew Kelly concludes his magisterial history of Dartmoor, *Quartz and Feldspar*, with the observation that the moor today is becoming 'increasingly bosky' – a lovely Devonian word meaning 'attractively bushy'.[43]

But the story of Lustleigh Cleave also offers a cautionary tale. In the eye of many commoners, what's happened to the Cleave is essentially land abandonment. Yes, grazing ceased almost by accident rather than because of calamity: through a combination of demographic shifts, bureaucratic inertia and the archaic quirks of English land law. And the Commoners Association has continued to draw an income from the public purse despite all of this. But for many farmers, seeing land taken out of production still hurts. It strikes deeply at a farmer's sense of pride and self-worth. It's one of the key reasons why rewilding is resisted so vociferously by some farming communities. If we're to persuade more farmers and common graziers to help restore our rainforests, we have to find a way to make such a project rewarding, both financially and as a moral cause.

Restoring Britain's rainforests cannot simply be left to accident. It may be that more commons go the way of Lustleigh in future; but I would far rather that the restoration of our rainforests becomes a cause that farmers and commoners participate in voluntarily, rather than out of resignation. We need to find ways to make this a proactive decision, in which commoners associations are incentivised to restore habitats, and where communities benefit from doing so. Brexit means the UK has now left the EU Common Agricultural Policy, with the government in the process of bringing in a new system of farm payments. Fortunately, at the time of writing, this new system seems to be heading in the right direction: underpinned by a principle of 'public money for public goods', in which farmers and landowners will be paid to restore nature. My hope is that the commoners of Lustleigh Cleave will benefit from this new system, too, perhaps as part of a wider 'landscape recovery' project to reconnect temperate rainforest fragments down Dartmoor's eastern edge.

Lustleigh's chequered history, however, shows us that this transition is unlikely to be an easy one. Commons are complex places, fraught with arcane politics. They remain a vital part of Britain's heritage, a link to a more communal past, and a reminder of the historic injustice of enclosure. But we mustn't let romanticism blind us to the ways that modern commons are also deeply flawed. Thanks in no small part to the blunders of past governments, the ideal of commons has become corrupted so that unsustainability and individualism seem hardwired into their design. By converting communal customs into individualised property rights, the 1965 Commons Registration Act unleashed the self-interest of commoners, pitching them into a war against one another and against nature.

This is not an atmosphere conducive to collaboration; quite the opposite. When speaking to Devon farmers, I've been told dark tales of commoners bullying and intimidating one another in disputes over grazing rights. On one Dartmoor common, the commoners fell out so badly that that they ended up taking one another to court. As a consequence, they've been unable to agree to a new Environmental Stewardship scheme for their common, with its attendant compensation payments for reducing livestock numbers. This, in turn, has led them to try to make up their lost income by grazing as many sheep on the land as their common grazing rights allow. The resulting over-grazing has turned the common into a lunar landscape, with Natural England assessing it as being in 'declining' condition.[44]

Unravelling this Gordian knot may prove intractable; but we have to try to find a way through. Some commons associations have been able to come to agreements over letting trees return to the landscape. At Holne Moor, on Dartmoor, progress is being made to re-wet some areas of the common and allow willows to regrow.[45] In Cumbria, the Ravenstonedale Common Graziers Association have agreed to plant a portion of their common with native broadleaved trees, fencing off the area with exclosures to keep out livestock.[46]

Whilst the extraordinary regeneration of Lustleigh Cleave may be unusual, discovering it's taken place has given me huge hope. It feels to me like an affirmation of the power of nature – and makes me believe there's cause for optimism for our other lost rainforests yet.

And hope springs from another source, too. In some parts of Britain, there are efforts to reinvent the commons for the twenty-first century, finding new ways of collectively owning and caring for land which puts ecological restoration at the heart of a new social contract. To observe what that looks like, we need to take a journey to the west coast of Scotland.

9

Forest People

Five figures stood huddled in the clearing: three women, two men, all of them wearing feather headdresses made from the plumage of tropical birds. Smoke coiled through the trees, rising from an ember held aloft by one of the men, his yellow tunic woven with bright geometric patterns. As the women shook rattles made from gourds, their faces daubed with ochre paint, one began to chant, her voice ululating through the forest. I watched, shivering, entranced by the spectacle of a traditional Amazonian ceremony. Except this event wasn't taking place in the Amazon, but in a Scottish rainforest.

The event had been organised by the Alliance for Scotland's Rainforest – a coalition of conservation charities, community groups and landowners – who had invited over representatives of Indigenous peoples from Brazil to lead a 'rainforest blessing' at the Cormonachan community woodlands in Argyll. Coinciding with the COP26 global climate talks taking place in Glasgow in November 2021, the ceremony was billed as 'a moment of reflection and solidarity' to highlight 'two very different rainforest habitats, profoundly linked by shared hopes and challenges'. Having spent the week hunched over laptops watching the interminable climate negotiations drag on, Louisa and I had been only too glad to escape to the woods that weekend to witness this special occasion.

The sacred blessing centred on the planting of a native oak tree within the forest – symbolising, according to the plaque planted next to it, 'our shared hopes for the future of rainforests in Scotland and the Amazon'. As the ceremony unfolded, a translator told the dozens of onlookers more about who the group were. They represented several different Indigenous communities, brought together by the Association Jiboiana, a Brazilian environmental NGO whose name is Portuguese for 'boa constrictor'. One of the women, Narubia Werreria, was an environmental activist from the Karajá people in the central Brazilian state of Tocantins.[1] Though the state lies within the legal boundaries of the Amazon, it's been almost completely deforested for soybean plantations and cattle ranching. Val Munduruku, meanwhile, was born and raised on the upper reaches of the Tapajós River in the northern state of Pará, and has campaigned against the illegal mining which spreads disease and destruction in her homeland.[2]

Witnessing the ceremony and hearing about the struggles of these Indigenous activists was deeply moving. It reawakened childhood memories of the 'Save the Rainforests' banner I'd painted as a kid, and the pictures I'd seen of the Kayapo people of Brazil leading some of the first protests against the destruction of the Amazon rainforest in the 1980s. That past week at the Glasgow climate talks, Indigenous peoples had won some long-overdue recognition from world leaders, with the announcement of a £1.25 billion fund for forest protection work led by Indigenous groups and local communities across the world.[3]

After all, Indigenous peoples comprise just 5 per cent of the world's population, yet look after some 80 per cent of global forest biodiversity within their territories.[4] Without a doubt, the most ecologically intact parts of the Amazon are those designated as Indigenous lands.[5] But Jair Bolsonaro, the far-right populist President of Brazil elected in 2018, has been waging an all-out assault on Indigenous rights and lands. Bolsonaro's foul-mouthed, racist tirades have frequently disparaged Indigenous peoples, claiming they are still 'evolving' and 'do not have culture', and voicing his disappointment that 'the Brazilian cavalry hasn't been as efficient as the Americans, who exterminated the Indians'. He has sought to dismantle protections on Indigenous lands in the Amazon, calling them an 'obstacle to agri-business'.[6] Rainforest deforestation rates have spiked under Bolsonaro's reign to their highest levels for fifteen years.[7]

The reverence and awe displayed by the audience at the rainforest blessing ceremony seemed a world away from the sneering condescension of President Bolsonaro. Yet as I watched, questions formed in my head. Weren't we, too, in danger of patronising our Amazonian guests in a more subtle way? Wasn't this just another instance of westerners exoticising Indigenous peoples, fascinated by the mystical rituals of the proverbial 'noble savage'? I began to wonder, too, what lessons we could realistically hope to draw from the event for our own broken relationship with nature. After all, who can claim to possess 'Indigenous wisdom' in modern Britain?

One possible answer presented itself in the form of a man with a ginger beard, who had stepped forward to take part in the ceremony. He was introduced as Àdhamh Ó Broin, a local resident whose ancestors had lived in the area for centuries. He stood facing the Indigenous delegation, his plain black sweater and blue jeans standing in stark contrast to the bright colours worn by Narubia and her compatriots. Only a small silver pendant hanging around Àdhamh's neck hinted at any sense of ritual. Then I noticed that he was standing barefoot on the damp, mossy earth, water squeezing out between his toes. He began to address the Brazilians in Gaelic, the native Celtic tongue of Scotland, classed as an indigenous language by the Scottish government.[8]

At first, the dialogue of Gaelic and Portuguese served to make the ceremony still more unfathomable to monoglots like me. But then something happened that required no translation. Àdhamh said something I couldn't make out, but which caused Narubia to weep. He then balled his hand into a fist and clasped it over his heart; a sign of empathy and solidarity, a statement beyond words that the struggle to save the rainforests was a shared cause. A collective whoop of defiance rose from the huddled group.

Later, I discovered that Àdhamh refers to himself publicly as an 'indigenous activist'.[9] I wasn't sure what to make of this at first: previously, the only white people I'd heard making claims to be 'indigenous' were fruitcakes and racists, shortly before they started complaining about 'bloody immigrants'. But I could find nothing resembling such bile in Àdhamh's public statements; not even a harsh word to say about the bullying English and their history of oppressing the Scots.

It soon dawned on me that others, too, have sought to construct modern Scottish identity around values of inclusion and ecological belonging, rather than an oppositional, exclusionary nativism. The Scottish writer and ecological campaigner Alastair McIntosh speaks unashamedly of having an 'indigenous childhood', raised on Celtic mythology amongst Gaelic speakers, living close to the land in a traditional crofting area. McIntosh has argued powerfully for reclaiming a sense of belonging to the land through reconnecting to 'indigenous histories and geographies', and that a 'Celtic spirituality . . . connecting soil, soul and society manifestly can and does exist'. This, he proposes, is not the basis for some tub-thumping nationalism, but is more simply about reclaiming a sense of community.[10] By participating in the blessing ceremony in a rainforest, I wondered whether Àdhamh Ó Broin was articulating a desire to reclaim Scotland as a rainforest nation.

It also struck me that – despite their obvious differences of scale and climate – Scotland and Brazil are locked in similar struggles over the fate of their rainforests. Both countries suffer from greatly unequal land ownership, with land concentrated in the hands of tiny elites. In Brazil, wealthy agribusiness barons control huge swathes of land, their vast estates, or *latifundia*, threatening the integrity of the rainforest. A recent study found that the biggest estates are concentrated in the 'arc of destruction' along the southern and eastern edges of the Amazon, where deforestation for soy plantations and cattle ranching continues to skyrocket.[11] In Scotland, fewer than 500 people own half of all privately owned land, according to the investigative campaigner Andy Wightman.[12] The remaining fragments of Scotland's temperate rainforests – though obviously far smaller than the tropical rainforests of Brazil – are similarly threatened by the ecological destruction wrought by these vast private estates, as we will later see.

With the ceremony coming to a close, and the chill November winds blowing in off the nearby Loch Goil causing everyone to shiver, we retreated to the local village of Lochgoilhead to warm up. Sitting in the timber-framed village hall, eating Tunnock's teacakes and drinking urn-boiled tea, our audience of around forty listened to speeches from the community group that manages Cormonachan Woods, from Members of the Scottish Parliament and from conservation NGOs.

Then Narubia Werreria took the podium. 'Indigenous peoples are the true guardians of the rainforest, its last line of defence,' she said, pausing to wait for the translator. Yet claiming an Indigenous right to belong, she continued, did not mean claiming ownership of the land in a western sense. 'We are not the owners of the Earth,' she said. 'She owns *us*. We need a relationship not of domination, but of mutual affection.' She paused again, and gestured to the audience. 'This domination begins in *your* own native forest.' Narubia's closing words made the air in the hall seem to crackle with electricity. 'We come here to offer an alliance. We have to unite, and this alliance must spread over all the Earth.'

Later that evening, her words reverberated in my head. I still had nagging questions about how Indigenous ideas of belonging to the land could be put into practice in modern-day Scotland, and doubts about who could legitimately claim to speak for a community. But freshly energised by this experience, I set out to better understand Scotland's rainforests, the threats they face and the people who live amongst them.

Scotland's rainforests are truly awe-inspiring places. The scale of the Scottish temperate rainforest zone, spanning the country's western Atlantic coast from Argyll in the south to Wester Ross in the north, is breathtaking. But what really puts Scotland's rainforests in a different league to others in Britain isn't just their spread; it's how unbelievably wet they are. For most of the year, they're simply drenched with rain. This means that a large part of Scotland's western coast has a climate that ecologists consider to be not merely *oceanic*, but *hyper-oceanic*.

It's a cliché to say that the Scots have numerous words for rain, but it's also true. I love how they embrace their torrential weather in this way, savouring the subtle differences in precipitation like a connoisseur of fine wines. *Dreich*, meaning dreary and dismal, was voted Scotland's favourite word the other year. *Smirr* refers to a fine drizzle, the sort of occult precipitation you get blowing in off the Atlantic. *Spindrift* is spray whipped up by the wind, whilst to describe rain as *stoating* means it's pouring down so hard, the raindrops are bouncing off the ground. But my favourite Scottish weather expression has to be *drookit*, meaning totally, utterly drenched.[13]

I got completely drookit on some of my visits to Scotland's rainforests, but that just comes with the territory. It's thanks to this volume of rainfall that Scottish rainforests support an incredible wealth of lichens, mosses and other epiphytic plants.

Species that are rare even in other British rainforests can be found teeming in superabundance in Scotland's wettest woods. And though many of the rainforests of England and Wales are dominated by oak and ash, in Scotland you also get rainforests of Scots pines – relics of the old Caledonian Wildwood – and, oldest and strangest of all, ancient groves of Atlantic hazel wood.

Louisa and I began our road trip around Scotland's rainforests in the aftermath of COP26. As part of her work as a campaigner with Greenpeace, Louisa had spent the preceding week cooped up inside the soulless conference centre where the Glasgow climate negotiations had been taking place, trying to nudge negotiators towards an ambitious deal. It was a relief for us both to break free from the suffocating world of climate diplomacy – with its endless prevarications, fiddling whilst the world burns – and breathe in the fresh air of some of the habitats we hoped to see saved. We drove the long, winding road that follows the edge of Loch Lomond; stopped to drink flask coffee and lap up the staggering scenery of Glen Coe; and checked into an Airbnb in Fort William, our base for a few days.

Each site we visited instantly brought home to me how rich Scotland's rainforests are. At Ariundle Oakwoods, north of Loch Sunart, we marvelled at the lungwort resplendent on every tree we passed. At Glen Nant, the trunks of trees had turned blue under a blanket of Lob scrob lichens. Some of the most spectacular rainforest trees occurred in unlikely places. Stopping for a pee by the side of a road, I chanced upon an extraordinary hazel whose bark had disappeared under the weight of epiphytes growing on it. I counted at least seven different species of rainforest lichens, clothing the hazel tree with a technicoloured dreamcoat, a riot of shades running from bruise purple to acid green.

One place, in particular, underlined the sheer vibrancy of life in Scottish rainforests. South of Fort William, on the shores of Loch Creran, is the Glasdrum National Nature Reserve. There we met Ben and Alison Averis, the two bryologists who'd shown us round Glen Coe, clad in their trademark waterproofs and looking as pleased with the miserable weather as ever. A sign at the mostly empty car park welcomed visitors to 'The hanging gardens of Glasdrum'. To my mind, this wasn't mere hyperbole: the rainforests here have to rank amongst the wonders of the world.

The very first stand of trees we came to were absolutely heaving with lichens. Tree lungwort plastered the trunks, arching its scaly green wings. Multiple different species of fish-smelling *Sticta*, shading from olive green to dark brown, competed with one another to gain the attention of our noses. And protruding from the bark of a willow was something surely unearthly: blue-black and gelatinous, seemingly watching us with hundreds of tiny eyes on stalks. I peered to get a closer look at this bug-eyed monster.

'That's a jelly lichen, called *Collema fasciculare*,' explained Ben, clearly overjoyed. 'Those octopus suckers are the fruiting bodies. You only really find a lot of it in western Scotland. It's like something out of a sci-fi movie, isn't it? *They've landed!*'

As we ventured deeper into the wood, a gentle smirr falling around us, the vegetation only got weirder. I'd grown used to seeing rainforest trees carpeted with mosses, but in Glasdrum many of the tree trunks seemed to have grown a new skin, so thickly smothered were they with scaly lichens. The bark of a hazel had been almost completely obscured by the riotous panoply of species growing on its surface. Satin-green *Lobaria virens*, the grey scallop shells of *Degelia atlantica*, the olivine foliage of *Sticta sylvatica*: they luxuriated in abundance. Purple filaments of tamarisk liverwort (*Frullania tamarisci*) ran like veins along the hazel's branches. Bright-green *Bazzania trilobata*, a liverwort with little legs, scampered down a boulder, delighting in the rain.

This damp, slimy, alien-sounding world may not be everyone's cup of tea – but to me, it was paradise. My hair was soaked, raindrops ran down my neck, and the pages of my notebook were so wet they'd started to gum together, but I didn't care. And though Glasdrum's trees seemed at times to be drowning under the weight of the epiphytes they supported, the lichens and bryophytes were in reality doing no harm to their hosts.

Yet a predator does stalk these rainforests. And, like any good predator, it's very hard to spot. Cloaked almost to the point of invisibility, it stalks its prey through the rain-soaked jungle. Its presence is betrayed by the carcasses of the victims that it leaves suspended in the branches. Fortunately, it's not a predator that would trouble Arnold Schwarzenegger. It is, you see, a fungus.

Crouched down in the undergrowth, Alison pointed to something in a nearby hazel tree. Several limbs of hazel had broken off and were hanging horizontally in the air, like a mobile. We edged closer. The horizontal branches were glued to the upright hazel staves with something oozing out of the bark, black as treacle. 'What *is* it?' asked Louisa in horror. 'It's called glue crust fungus,' explained Alison. 'It feeds on dead hazel wood by catching branches as they fall with this sticky black glue. Isn't it amazing?'[14] The 'hanging gardens' of Glasdrum were certainly living up to their name.

Death begets new life in the rainforests in other ways, too. We came across the fallen trunks of two oaks, casualties of a past storm. But though these veteran trees had been knocked down, they were getting up again. What had once been horizontal branches were now the vertical trunks of resurgent trees, sending out fresh twigs, putting down new roots, readjusting to the ninety-degree change in their fortunes. The root balls of the upturned oaks had become earthen banks at right angles to the wind, providing shelter to fresh saplings which were now bursting forth from the forest floor.

'They're called phoenix trees,' said Ben. 'New life rising from the ashes of the old.'

'You can't kill a tree,' said Alison, before adding: 'Unless you're a human, that is.'

Despite the vitality of many of Scotland's rainforests, the overall picture is not good. Only 74,000 acres (30,000 hectares) of Scotland's rainforests remain. That's just a fifth of the area that has the climatic conditions suitable for this habitat to thrive.[15] The rainforests Louisa and I visited were mainly low-lying sites, hugging close to the shores of lochs and Scotland's Atlantic coastline. But whenever we raised our sights to the horizon, it was plain to see how bare the hillsides were. The flayed land rippled like exposed muscle.

Some of the threats faced by Scotland's temperate rainforests are familiar to those faced across the rest of Britain: the problems of overgrazing by sheep, destruction by plantation forestry, and habitat fragmentation. But two other factors in particular imperil Scottish rainforests, in ways that are felt less acutely elsewhere: the twin menaces of deer and rhododendron.

Deer populations in Scotland have reached epidemic proportions, compounding the problem of overgrazing by livestock. The Alliance for Scotland's Rainforest estimates that deer are now responsible for around 80 per cent of the overgrazing pressures in Scottish rainforest sites.[16] *Rhododendron ponticum*, meanwhile, is an invasive species of plant whose rampant spread threatens to choke the life out of the nation's rainforests. Neither of these threats, however, are natural in origin: instead, they're the result of human interference in the landscape. Each of these rainforest killers has arisen from Scotland's history of vastly unequal land ownership and exploitation.

Their story begins with the Clearances: the forced eviction of thousands of inhabitants of the Highlands and western islands in the late eighteenth and early nineteenth centuries.

Following the bloody defeat at Culloden of the army of 'Bonnie Prince Charlie', the Stuart Pretender to the British throne, Highland culture came under assault from the British state. The clan system that had dominated the Highlands, binding together communities in kinship groups, was deliberately broken. Clan tartans were outlawed, the Gaelic language suppressed, and landowners who had supported the Pretender forfeited their estates. This paved the way for wealthy British aristocrats to buy up huge swathes of the Highlands, who proceeded to evict their tenants from the land in the name of 'agricultural improvement'.

By the late nineteenth century, just fifteen landowners owned half of the Highlands; the Duke of Sutherland alone possessed over a million acres.[17] Small-scale farms were replaced by huge sheep ranches, stocked with a hardy breed called the Cheviot that could survive in the harsh, wet climate of the Highlands. The historian John Prebble recounts how the coming of the Cheviot presaged doom for many tenant farmers: 'word of it reached the Highlands in the old way, on the lips of a seer who travelled from township to township, calling a warning . . . "Woe to thee, oh land, the Great Sheep is coming!"'[18]

Still worse was to come, however. The potato famine of the 1840s further ravaged rural communities, and accelerated emigration from the Highlands to the colonies. Later, increased imports of cheap lamb and mutton from Australia undercut domestic production, making even the Cheviot unprofitable across swathes of Scotland's most marginal uplands. But the decline of sheep farming in fact merely opened up the land to different forms of exploitation. Having almost emptied the Highlands of people, the landowners in possession of it now decided to fill the land with deer, grouse and pheasants.

During the Victorian period, the Scottish Highlands became, in the words of Andy Wightman, 'a vast outdoor playground for the upper strata of British society'.[19] The aristocratic fashion for deer-stalking and grouse-shooting was first sparked by Queen Victoria and Prince Albert's purchase in 1852 of the Balmoral estate in the Cairngorms. Suddenly, the English upper classes were seized by the new 'Balmorality'. 'Everybody who was anybody in 1850 wanted a Highland sporting estate,' writes one historian.[20] Deer-stalking wasn't done merely to dine off venison: it was about prestige. The sport's appeal was epitomised in Edwin Landseer's famous painting *The Monarch of the Glen*, depicting a stag whose magnificent antlers many a deer-stalker yearned to bag for their trophy wall. The Scottish theologian and Gaelic enthusiast Thomas MacLauchlan voiced concerns that Victoria and Albert's penchant for Scotland 'may hasten the consummation of making our Highlands a great deer-forest by inducing a larger number of our English aristocracy to flock to them for the purpose of sport'.[21]

MacLauchlan's fears were soon realised. At the start of the nineteenth century, there were barely seven deer forests in Scotland actively managed for hunting; yet by the century's close, as many as 150 of them sprawled over a staggering 2.5 million acres of the Highlands.[22] Nor is this a fad that has since faded. As the aristocracy have declined, the fashion for buying up Scottish sporting estates has been taken up with gusto by City bankers, Russian oligarchs and Middle Eastern sheikhs. At the start of the twenty-first century, the combined area of Scotland's deer forests and grouse moors was estimated to run to around 5 million acres: around a quarter of the Scottish landmass.[23]

The result has been an ecological disaster. You might think an increase in deer-stalking would result in fewer animals, but sporting estates have an obvious interest in maintaining high numbers of red deer: large herds make for easy sport. Roe deer populations, too, have boomed. And to make matters worse, some sporting estates historically introduced non-native deer species – sika and fallow – and allowed them to escape into the wild. Deer can leap over all but the highest fences; and with no natural predators like wolves to control their numbers, culling remains the only real check. Overall numbers of deer in Scotland have thus skyrocketed, rising from around 100,000 after the Second World War to perhaps as many as 1 million today.[24] This population explosion has ramped up pressure on the woods in which deer browse.

Repeated warnings about the devastating impacts of increasing deer numbers had fallen on deaf ears. The respected ecologist Frank Fraser Darling cautioned as long ago as the 1950s that deer numbers in Scotland were unsustainable, suggesting a cull to reduce populations to around 60,000 animals. In 1959, a statutory Red Deer Commission was set up to advise on the problem, but was granted too few powers to enforce any real action. Reading through extracts from a half-century of reports reveals the Commission's growing exasperation at the failure of private sporting estates to get a grip on deer numbers. The late Deer Commissioner Simon Pepper observed how the organisation's tone gradually shifted from 'mild advice, to frustrated pleading, to warnings and finally direct threats of legislative change' – all, alas, in vain.[25]

Much of this burgeoning deer population has taken shelter in Scotland's forestry plantations, which expanded hugely in the same period. But marauding deer don't confine themselves only to plantations: they've also invaded Scotland's remaining native woods, including its fragments of rainforest.

The impacts are painful to behold. Ancient woods wrecked by deer overbrowsing look dead on their feet; veteran trees stand stricken and lonely on barren hillsides that used to be forests, watching their remaining companions collapse into senescence. James, an ecologist who tweets as @Collbradan (Irish Gaelic for 'hazel-salmon'), discovered one such site in the glens above Fort William: a place that on Google Earth appears totally devoid of trees. But after examining old maps, James realised that this glen had once supported several wild woods, each named in Gaelic: woods such as *Cam Dhoire*, meaning 'the crooked thicket', or *Brian Choille*, 'rough wood'. Yet, when he visited the area, only a few scattered surviving trees, creaking with age, attested to the ecosystem that had once carpeted these hills.

Why? Because, as James explains, 'each tree that died simply wasn't replaced, as every seedling that tried to grow up over the past 150 years has been eaten by sheep or deer'. The people who had once inhabited this landscape and given names to the woods had long been cleared away, allowing 'centuries of decline to go unnoticed'. Consumed by deer and forgotten by humans, these ghosts of past rainforests linger on; but without a new generation of trees to replace them, perhaps for not much longer.[26]

So what can be done about Scotland's deer problem? In search of answers, I spoke to the conservationist Gordon Gray Stephens, who has worked on the management of Scotland's native woodlands for decades. I'd first met Gordon at the rainforest blessing ceremony at Cormonachan, where he'd stood out from a mile off. Dressed in kilt and tweeds, he'd been wearing a sporran made out of a dead badger, its glass eyes staring beadily. Over tea, I'd sidled up to Gordon to ask how the badger had ended up where it was, and was relieved to hear his father had made it from roadkill. Later, I gave Gordon a call to ask him about deer.

'The politics of deer management is very vexed,' he told me. 'With livestock, such as sheep, they're at least controllable. Farmers have an economic interest in livestock and a legal responsibility for them. But with deer – they're not actually owned by anyone.' Sporting estates, he explained, don't own the deer themselves, just the right to shoot them when they come on to their land. As a result, no one really has ownership of the problem, and landowners are rarely legally required to cull deer. The Scottish government's environment watchdog, NatureScot, has never made use of the powers available to it to introduce a statutory control scheme, relying instead on estates entering into voluntary agreements.[27]

What's more, whilst sporting estates have an incentive to kill red stags – the ones with the prized antlers – there's no similar sporting motivation to control roe deer, or the invasive sika. 'Roe deer are often just ignored,' Gordon told me. 'To a certain type of deer manager, they just "don't count".' By way of illustration, Gordon sent me the deer management plan drawn up by a group of estates who together own most of the Morvern peninsula. Morvern lies in the wet heart of Scotland's hyperoceanic zone, and boasts some spectacular fragments of temperate rainforest. Yet the plan for managing its deer focuses almost exclusively on red deer, with no policy in place for controlling roe deer numbers. 'A number of Estates have significant populations of roe deer,' the plan says, admitting: 'Roe are capable of causing very significant damage both in commercial and amenity woodland areas.'[28] Yet, Gordon sighs, 'most estates here don't even recognise there's an issue'.

Fundamentally, Scotland's rainforests will remain threatened whilst deer numbers remain so high. 'Oak seedlings start to disappear once you get two to four deer per square kilometre of woodland,' Gordon explained. 'And there are far higher deer densities than this at present.' Across Scotland's publicly owned forestry estate, population densities are now more like twelve deer per square kilometre.[29]

Having allowed deer populations to balloon to such unnaturally high levels, it should surely be incumbent on sporting estates to help fix the problem. In 2020, a coalition of Scottish environmental groups called for deer culls to be made compulsory for landowners.[30] This is, after all, the norm in other parts of Europe, where deer numbers are much more closely regulated and governments set statutory targets for culls.

A report published later that year by the independent Deer Working Group, commissioned by the Scottish government and chaired by Simon Pepper, stopped short of setting compulsory cull targets for landowners. But it did make other far-reaching proposals, including changing the law to give NatureScot greater powers to step in and cull deer where they're causing damage. Politics is the art of the possible, and Pepper knew his report would have to overcome deeply entrenched vested interests opposed to any outside interference in sporting estates. In terms of getting political buy-in, it seems to have worked: following the 2021 Holyrood elections, the power-sharing deal signed by the Scottish National Party and the Scottish Greens agreed to implement virtually all of the Deer Working Group's recommendations.[31] Whether this will be sufficient to significantly reduce deer numbers, however, remains unclear.

What's certain, however, is that without greater control over deer populations, Scottish rainforests will continue to decline. The Alliance for Scotland's Rainforest warns that browsing pressures in many rainforest sites are 'limiting their long-term survival'.[32] And deer are not the only threat to the rainforests arising from large sporting estates. A second, even deadlier menace lurks in the undergrowth, hiding innocently behind beguiling purple flowers.

The ancient hazel woods of Ballachuan are barely half-an-hour's drive south from Oban. But to Louisa and me, it felt like we were skirting the edge of the known world. Our destination lay on the Isle of Seil, sundered from mainland Scotland save for a humpbacked stone arch known as the 'Bridge over the Atlantic'. A mist had blown in off the ocean, cloaking the land in a milky haze, through which the sun shone like a weak and distant star. The island's main road led us to a lonely church, built inexplicably in the middle of nowhere. Between us and the sea, the land arched its back, forming a long, rocky ridge, like the spine of a whale carcass washed up on the shore.

Clinging to this outcrop was a wood: a tangle of spindly limbs, bent double against the weather, with leaves the colour of faded sunlight. This was no ordinary wood. It was composed almost entirely of hazel trees, and it may have stood on this spot since the last Ice Age. Ballachuan is a prime example of Atlantic hazel wood, a flavour of rainforest almost unique to Scotland – but which would once, long ago, have stretched all the way down the western coast of Britain.

Here to show us around was Ian Dow, woodland coordinator for a local environmental charity, the Argyll and the Isles Coast and Countryside Trust, or ACT for short. Ian was bearded and wore his hair in a ponytail; his khaki anorak and sturdy leather boots spoke to a life lived outdoors. As we headed towards Ballachuan, I asked him how he came to have such an awesome job. Ian told me how he'd started out as a tree surgeon, but didn't like being ordered by councils to chop down too many healthy trees. So he'd thrown in the towel and become a community forester instead, moving from Lincolnshire to live on the community-owned Knoydart Estate on Scotland's west coast. More recently, he'd joined ACT to lead their Saving Argyll's Rainforests project, combining practical restoration work with engaging communities to help tackle the threats they face.

We walked together over meadows teeming with waxcap fungi, bright reds and yellows peppering the grassy sward, a sure sign that these fields had been left unploughed for centuries. Everything here felt ancient, almost primeval: this was land that had been left to its own devices, given time to grow rich and complicated.

To enter Ballachuan, we clambered over a stile and ducked beneath a clatter of branches. The forest floor was a gorgeous mosaic of hazel leaves, shading from lemon-yellow to burnt umber. I stared up at the trees from which they'd fallen. Each hazel formed its own thicket, a cluster of boughs arcing upwards. But they were nothing like the hazel trees I was used to seeing, all thin coppiced stems, ramrod straight: these veterans were ancient and gnarly, with thick, crooked trunks. They stooped like old men, their weather-beaten faces grizzled with mossy beards.

The wood's canopy of wind-sculpted twigs, though not quite as low as the stunted oaks of the Dizzard or Young Wood, nevertheless allowed in plenty of light. Each hazel, as a result, played host to a rambunctious collection of lichens. Lime-green *Lobaria virens* bloomed on the sodden branches, jostling with an efflorescence of the grey scallop-shelled *Degelia atlantica*, which stuck to the trees like a coating of barnacles. Ian got out a hand lens to peer closer. Even on the few areas of exposed bark, the hazels were supporting a menagerie of camouflaged species. Tiny script lichens, their fruiting bodies a succession of black dots and dashes, pebbledashed the branches with their indecipherable language. Louisa and I gently ran our fingers over the bark, trying to read the lichens like Braille.

We tend to think of hazel as merely a shrub of the understorey, too easily overlooked in woods dominated by taller trees of oak and ash. But Ballachuan gives the lie to such condescension, revealing that hazel is quite capable of forming woodlands all by itself. It's only recently that Atlantic hazelwood has become recognised as a very special habitat in its own right, thanks in large part to the work of the botanists Sandy and Brian Coppins. 'This habitat has been with us for thousands of years,' they wrote a decade ago. 'But surprisingly little is known or written about Atlantic hazelwoods, and as a habitat it receives very little attention, or even recognition.'[33] The richness of the lichen and bryophyte flora of Atlantic hazelwood puts it firmly within Britain's temperate rainforest biome. And whilst there are tantalising signs that fragments of it may exist in Cornwall and Wales, its heartland is undoubtedly western Scotland.[34]

This is all the more surprising when you consider the long history of people using and revering hazel trees. Hazelnuts were an important source of protein for people living in Scotland as far back as the Mesolithic. Hazels are also associated with knowledge and creativity in Celtic mythology: the legendary warrior-queen of Skye, Scáthach, is said to have acquired the wisdom to make peace after consuming roasted hazelnuts. In old Gaelic, hazel is *coll*, or *calltain* in the modern tongue. Old place names hint at former times when Scottish hazel woods were perhaps better known, such as *Camas a' Challtainn*, 'the bay of the hazel', on the rain-soaked shores of Loch Shiel.[35]

Hazels aren't much good for timber, but their fast-growing stems have been used throughout human history for fencing and firewood, and for the wattle-and-daub walls of medieval buildings.[36] In order to harvest hazel sustainably, our ancestors developed coppicing, a practice with which hazel has become so synonymous that we tend to assume all hazel trees have been managed this way. But Ballachuan, which likely hasn't been coppiced in centuries, reminds us that hazels send up basal shoots naturally, and probably evolved this mechanism to cope with browsing by herbivores. And though there's a well-meaning desire to rejuvenate traditional coppicing practices in many 'neglected' woods, coppicing old-growth hazel supporting so many lichens and bryophytes would be a very bad idea. As Sandy and Brian Coppins write, 'The risk of allowing coppicing in this habitat is that all the accumulated wildlife interest will be eradicated, and the re-growth will simply be a barren set of stems.'[37]

The hazel stems at Ballachuan were anything but barren. Ian, Louisa and I moved from tree to tree in rapt fascination, calling out to one other each time we found some new delight. Ian's particular passion, he told us, was for fungi: the weirder, the better. 'It was the symbiotic and parasitic stuff that really drew me into fungi and lichens,' he said, with a geekish grin, and pointed towards the hazel branch hanging in front of us. It was stuck together with the pitch-black gum of glue crust fungus which we'd now grown to recognise since our earlier encounter with it at Glasdrum. But the specimen that Ian had found was displaying something stranger still. Wrapped around the branch was what looked like a small orange hand.

'This is hazel gloves fungus,' Ian beamed proudly. 'It feeds parasitically on the glue crust fungus, which feeds on the decaying hazel wood.' We leaned closer to get a good look at this miniature horror, its saffron-coloured fingers clasping tightly to the tree, like a child's mitten or a tiny baseball glove. 'When they first emerge, they can be bright orange,' Ian went on. 'Like Cheetos. Or Donald Trump's tiny hands.'

Hazel gloves fungus is vanishingly rare, and found exclusively in temperate rainforest. As the name suggests, it thrives particularly on hazel; western Scotland certainly has some of the best sites for it, although it's also been found in Wales, north Devon and increasingly in Cornwall. I was fascinated by how such a thing could even exist; how long it must have taken for this delicate, precarious food web to evolve. In Britain's rainforests, it's a fungus-eat-fungus world.

But this pinnacle of rainforest evolution also points to one of our rainforests' foremost threats. The Latin name for hazel gloves fungus is *Hypocreopsis rhododendri*, so-called because it was first found growing in North America on a rhododendron bush. But, for unknown reasons, the fungus seems not to prey on rhododendron on this side of the Atlantic. Which is unfortunate – because Britain is now suffering from a plague of rhododendron that threatens to choke our rainforests to death.

As we left Ballachuan in search of a warming cup of tea, an invading rhododendron leered from a nearby roadside verge. For thousands of years, Ballachuan's hazel woods have survived alongside humanity. But now, as a result of recent human activities, they're under attack from a new and deadly foe. The rhododendron's leaves waved in the breeze. It seemed to be taunting us, as if it were saying: *we're coming; nothing is sacred; nowhere is safe.*

When the enterprising Dutch plant collector Conrad Loddiges first brought *Rhododendron ponticum* to Britain in 1763, he probably thought he was doing the country a favour. Here was a beautiful, exotic plant with magnificent purple blooms, guaranteed to brighten up drab gardens. But unwittingly, he had unleashed a monster.

At first, nothing seemed untoward. The first rhododendron bushes are said to have been grown by the Marquess of Rockingham at his stately home in Northamptonshire. Some specimens were germinated at the Royal Botanic Gardens in Kew. For a while, it remained a rarity, confined to a few landscaped gardens.

Yet the killer was simply biding its time. Rhododendron is an evergreen shrub that keeps its waxy, emerald-coloured leaves all year round. This makes it excellent cover for game birds like pheasant and woodcock to shelter under. As a result, rhododendron became increasingly popular with aristocratic sporting estates during the nineteenth century. A paper by the Forestry Commission notes that rhododendron was 'planted extensively in Victorian hunting estates . . . to provide shelter for game species'. Ominously, the most extensive plantings took place 'particularly in western coastal areas, under woodland canopies'.[38]

In fact, rhododendron flourished in exactly the same wet, mild parts of the country that support Britain's temperate rainforests. It thrived in the damp climate of Atlantic woodlands, and it received a helping hand establishing itself from the nation's gamekeepers. 'Suddenly, nothing but *R. ponticum* would do for game cover, and everyone wanted to get hold of it,' writes the botanist Richard Milne.[39] Had it just stayed where people deliberately planted it, rhododendron might not have become quite such a problem. But it turned out to be an aggressively invasive species, rapidly colonising woods, outcompeting other understorey plants and shading out young saplings with its evergreen foliage. What's more, by allowing deer and sheep to overgraze our western rainforests, we'd created even better conditions for rhododendron to spread. 'Overgrazing . . . churns up the soil, creating seeding establishment sites, which help it to invade,' notes Milne.[40]

Rhododendron ponticum spreads in two ways: through seed dispersal, and through what's called 'stem layering' – where branches touch the soil, they put down roots and produce clones. This leads to rhododendron bushes often becoming colossal in scale as they expand upwards and outwards: one was once recorded as measuring twenty-four metres in diameter.[41] Some rhododendron infestations have become so vast and impenetrable that hikers have been known to get lost in them and require rescuing. These alien invaders are real-life triffids. And rather like the carnivorous plants in John Wyndham's sci-fi classic, they're rampaging through the British countryside, wreaking havoc on our rainforests.

By the time ecologists had started to raise the alarm about rhododendron in the 1920s, it was too late to easily eradicate it from British shores. It had already established itself in wet woodlands from Cornwall to Snowdonia. Today, the Forestry Commission estimates that rhododendron covers nearly 250,000 acres (100,000 hectares) of Britain: an area around four times the size of Birmingham.[42]

The biggest and deadliest infestations of rhododendron, however, lie in western Scotland. Many Victorian sporting estates in Scotland supplemented their upland deer forests with lowland pheasant and woodcock shoots and formal gardens stocked with rhododendron. Whilst deer multiplied on the moors and marauded into woods, gamekeepers planted up the glen woodlands with invasive game cover. Together, these serial failures have left Scotland's rainforests facing a grim legacy. Out of the 74,000 acres of temperate rainforest left in Scotland, around 40 per cent is estimated to be infested with rhododendron.[43] NatureScot and Scottish Forestry have mapped a set of priority control areas for eliminating rhododendron, including the extraordinary rainforests we visited at Glasdrum, at Glen Nant and around Loch Sunart.[44]

Yet some sporting estates remain stuck in the past on this issue. The Isle of Islay, off Scotland's west coast, has a number of ancient woodlands that have been invaded by rhododendron: a result of one of the island's sporting estates historically planting it for game cover. ACT, the charity that Ian Dow works for, has been working with the island's landowners to root out the invasive species. But I was told of at least one sporting estate on Islay whose absentee laird was refusing to take part in the eradication programme because the rhododendron was providing good cover for the woodcock they enjoyed shooting in their woods.[45]

Removing rhododendron is hard enough even where landowners are willing to pitch in. 'Rhodie-bashing', as it's called, is a laborious, exhausting process. Pulling out small saplings is easy enough, but cutting down fully grown bushes is backbreaking work. Rhododendron can be flailed and mulched mechanically; but on steep ground, manual removal may be the only option. The woody brash generated often has to be removed: if it's left *in situ* to dry out, it can become a fire hazard, with piles of dead rhododendron brash said to have contributed to the devastating wildfires that ripped through the temperate rainforests in Ireland's Killarney National Park in 2021.[46] It doesn't help that removal necessitates a long-term effort, requiring conservationists to return to the same site over and over again. Nor is it enough to simply hack off the stems: rhododendron will regrow unless the roots are dug out, or the stumps poisoned with weedkiller. I hate the indiscriminate use of herbicides in the countryside, but to defeat a monster like rhododendron, their targeted use here seems justified.[47]

Given how difficult rhododendron is to eliminate, a starting point for the long-term management of this invasive species would be to ensure more of it isn't spreading into woods in the first place. And whilst it's nowadays illegal to plant *Rhododendron ponticum* in the wild, it is, bizarrely, still perfectly legal to sell it. That means it's still being bought, planted in gardens and self-seeding over a wider area. As we drove through Scotland's rainforest zone, we passed by a sign for a garden centre. My eyes widened in disbelief as I saw it: the centre was proudly advertising its presence with a huge picture of a purple-flowered rhododendron. Gardeners might love its blooms, but continuing to sell this invasive species in the heart of the rainforest zone is deeply irresponsible. Ideally, sales of *Rhododendron ponticum* should be banned entirely – after all, other non-invasive species of rhododendron exist, as well as countless other flowering shrubs, both native and exotic.

Some landowners are rising to the challenge of extirpating this threat. One example is on the Morvern peninsula, just to the south of Loch Sunart, which harbours many Atlantic oakwoods teeming with rare species. Morvern has geography on its side: the Alliance for Scotland's Rainforest calls the penin-sula 'one of the most defensible and practical areas in Scotland for large scale rhododendron eradication'.[48] It's also lucky to have some forward-thinking landowners prepared to take up arms against the invading rhodies – such as the Ardtornish Estate, owned by the Raven family; and RSPB Scotland, who are looking to establish a new reserve at Glencripesdale. Unfortunately, Morvern's rhodie removal project has struggled to raise funds: there have been three fundraising attempts since 2016, and it's still £2.6 million short of its anticipated costs. As the project overseers write: 'The more the project is delayed, the worse the rhododendron problem becomes, and the more expensive it will be to restore this globally important habitat.'[49]

Ultimately, rhododendron eradication needs more than just a few enlightened landowners and philanthropists. It's a mission that requires government intervention to ensure all landowners have to step up. Landowners are entitled to public grants to help pay for rhodie removal.[50] But the total costs of eliminating rhododendron from the quarter of a million acres over which it's now spread are projected to be astronomical. Scottish Forestry estimates that eradicating it from Scotland alone could cost £400 million.[51] Studies into the costs of removing it from Snowdonia National Park in Wales suggest anywhere between £10 million and £45 million.[52] Add in the rest of Wales and the Westcountry, and Britain could be facing an eye-watering £1 billion bill to remove a species first introduced because aristocrats thought it looked pretty and found it useful for pheasant shoots.

Getting all of this funded from the public purse seems unlikely: the Scottish Exchequer and UK Treasury would surely baulk at such a request. And given that it was big private estates that saddled us with this problem in the first place, why shouldn't those same estates be picking up the tab to fix it? Environmental watchdogs do have powers called 'species control orders' to compel landowners to remove invasive species like rhododendron, but these are almost never used. When I asked Ian Dow for his view, he was adamant. 'Of course all landowners should be strongly encouraged to remove rhodies,' he said.

The Scottish poet Rabbie Burns once bemoaned the loss of old woods at Drumlanrig in Dumfriesshire, mourning the 'stately oaks' with their 'twisted arms'. Were the trees destroyed by a storm, Burns wondered, or by some disease? No, he discovers: the true culprit was the landowner, the Duke of Queensberry. 'The worm that gnaw'd my bonie [bonny] trees, / That reptile wears a ducal crown.'[53] But if the big landowners won't act, what about changing who owns the land? What if we all had a greater stake in owning and managing the rainforests?

On Argyll's southern coast, not far from the Isle of Bute, lies the Cowal peninsula. A quick peruse of Andy Wightman's Who Owns Scotland website tells you that most of the peninsula's moorlands and forestry plantations lie in the hands of various baronets, absentee lairds and shady offshore companies.[54] But at Cowal's southernmost tip lies something more interesting: a sweep of Atlantic rainforest called Glenan Wood, owned and managed by the local community.

My plans to visit Glenan Wood in early 2022 were scuppered by Storm Eunice, whose fierce winds brought more than just a little smirr to western Britain. So I arranged instead to speak on the phone to Eve MacFarlane, a resident of the nearby village of Portavadie, and director of a community group called Friends of Glenan Wood.

'Glenan Wood's always been a very special place to me,' Eve told me. 'It's on a beautiful rocky coastline, where the woods roll down to meet the sea. The oaks are so gnarled and old, really mossy and covered in lichens. The woods smell sweetly of bog myrtle, and burns tumble down through them. It's a very wild and sensory place.' Eve's evocative descriptions told me everything I needed to know about her love for the wood.

'We've had a long family connection to Glenan Wood,' she continued. 'My great-grandma used to go camping there at the turn of the last century, and I remember running wild in the woods as a child.' After living in England for a while, Eve moved back to Cowal some years ago and set up a coffee-roasting business and café. Then, in 2016, Forestry & Land Scotland – up until that point the owner of Glenan Wood – announced it was up for sale.

'When I saw it was on the market, we only had two days to

put in a bid showing our interest,' remembers Eve. She recalled local residents gathering to discuss the sale at a neighbour's house, with lengthy conversations about what to do. 'The wood's such a beautiful place, there were a lot of people in the local community driven by a sense of wanting to protect it.' Her account brought back memories for me of my mad dash to find buyers for High Wood in Cornwall shortly before bids closed. But the newly formed Friends of Glenan Wood had Scottish law on their side.

In 2003, the recently devolved Scottish government brought in a Community Right to Buy. This powerful piece of legislation gives community groups first dibs when some local land comes on to the market. If the community group expresses an interest in purchasing it, the ordinary sales process is paused, giving time for the group to raise funds. This breathing space allows residents to set up crowdfunders, bid for grants, and draw up business plans for how the land would be used by the community in future. A Scottish Land Fund also allows communities to draw on public funding for buyouts. At the time of writing, more than half a million acres of Scotland have been bought by community groups, rejuvenating local economies and reversing some of the rural depopulation that came with the Clearances.[55]

At Glenan Wood, Scotland's Community Right to Buy law came to the rescue of a rainforest. 'That absolutely helped,' Eve told me. 'It gave us an incredible feeling of power.' It was still a long slog to complete the buyout: three years of campaigning, fundraising, and consulting with everyone in the local area. But, in Eve's eyes, it was absolutely worth it: 'So much private development is just done *to* communities, with little control over it. But, for this, the community can now decide on the future of the land themselves.' In 2019, Friends of Glenan Wood took ownership of 360 acres of temperate rainforest.[56]

The wood's former owners, the Scottish government body

Forestry & Land Scotland, had somewhat neglected it – perhaps seeing this remote pocket of ancient woodland as peripheral to its core business of managing commercial forestry plantations. But the new community owners employed a forest ranger who got straight to work, organising groups of volunteers to root out invasive rhododendron, and removing non-native conifers shading out the old oaks. Deer fencing has been erected to reduce overgrazing and encourage fresh sapling growth. Local schoolkids have taken part in a 'bioblitz' day to learn about the rainforest botany. And an old plantation of larch has been felled, with a community orchard planted in its stead. Future plans include making the wood more accessible to visitors, and potentially investing in some glamping pods to generate a fresh income stream for the community. 'The power of it being community owned is that we can focus on things like this,' said Eve.

Over the past twenty years, Scotland has seen a big expansion in the number of woodlands owned or managed by communities. Cormonachan Woods, for example – which we visited for the rainforest blessing ceremony at the start of this chapter – has been formally managed by a community group since 2006, though the land still belongs to Forestry & Land Scotland.[57] Just up the road from Glenan Wood, meanwhile, is another community-owned woodland. Kilfinan is a commercial forestry plantation, in which plots have been let out to new woodland crofts, giving more people the chance to make a living from forestry.[58]

The community ethos has also permeated other rainforest conservation efforts. Woodland Trust Scotland, for example, has recently begun a volunteer project to collect tree seeds from west coast rainforests, and germinate the seedlings in a tree nursery. It aims to fill a gap in the supply of local saplings – reducing reliance on imported trees, with all the attendant risks of bringing in new tree diseases – and has so far attracted support from 200 volunteers in its first year.[59] In Glen Creran, north of the 'hanging gardens' of Glasdrum, Gordon Gray Stephens and ACT both helped galvanise a community into eradicating rhododendron from the valley, with even keen gardeners uprooting it from their flower beds. Not only did this rid a rainforest heartland of an invasive species, it also brought the community together, generating new friendships in the process.[60] A report by Gordon on rainforest restoration concluded that community involvement has 'a fundamental impact on the sustainable removal of invasive rhododendron'.[61]

Hearing about all these projects was a breath of fresh air to me. Living in England, I'm used to a more deferential culture of cap-doffing towards big landowners, in which even our remaining commons have become corrupted cliques. We desperately need Scotland's Community Right to Buy laws to be replicated in England and Wales, and give more communities the chance to genuinely 'take back control' over the land they need and love.

Having a thriving community ownership sector, however, doesn't mean there are no arguments in Scotland about how best to use the land. Over the past few years, there have been rising tensions between some Scottish land reformers and proponents of the growing fashion for rewilding. In some quarters, rewilding is increasingly seen as the pastime of wealthy toffs, foisted upon local communities by a new wave of 'green lairds'. Some fear that big estates are using rewilding to greenwash their reputations or, worse still, enact a modern version of the Clearances, swapping sheep and deer for wolves and lynx. Given the history of rural depopulation in Scotland, it's understandable that some worry ecological restoration without popular consent could simply mean eco-colonialism.[62] Others point out that it's perfectly possible to make rewilding and land reform go hand-in-hand – pointing to examples like the Langholm community buyout, which is in the process of rewilding a grouse moor that formerly belonged to the Duke of Buccleuch.

In the run-up to COP26, these debates spilled over into a minor scuffle about Scotland's rainforests. The @DailyGael, an English–Gaelic Twitter account with a caustically satirical sense of humour, seized on a post by the Scottish Green Party about rainforests on Scotland's west coast. 'Focan' rainforest cove . . . the Greens are out of their tree,' they fulminated. 'Just when you think rewilding narratives couldn't get any more craicte and mythical' – *craicte* being Gaelic for 'crazy'.[63]

But the tweet ended up getting ratioed with a string of responses from Gaelic and English speakers alike – pointing out that, actually, Scotland *does* have rainforest; that it's a recognised habitat, not some mystical invention; and that it's something many in Scotland feel justifiably proud about. I saw that pride in Gaelic speaker Àdhamh Ó Broin's face when he took part in the rainforest blessing ceremony at Cormonachan, making common cause with Indigenous activists from the Amazon.

What I hadn't realised when I witnessed that meeting was that the term 'Gaelic' itself derives from a word meaning 'forest people'.[64] The Clearances not only removed people from the land; they sundered their connection to nature. The landowners who expelled them introduced ecological monocultures in their place, for profit and sport. Sheep, deer and rhododendron ravaged many of the surviving rainforests, erasing places once cherished by the Gaelic-speaking peoples who lived in their midst. Given such a heritage, restoring Scotland's rainforests seems less like an exotic imposition and more a mission of national reclamation.

And I'd argue that restoring our rainforests, far from being a form of 'eco-colonialism', is actually a decolonial endeavour. After all, we can hardly expect Brazil to protect its rainforests, yet hypocritically do nothing to bring back our own. We can't simply place all the burden of reversing biodiversity loss on the developing world.

But to have any hope of success, rainforest restoration has to be done with people at its heart. If we're to bring back our lost rainforests, it'll prove impossible to do so without the active engagement of the communities who live in and around them: the people who will walk in them, root out invasive species, mend fences, cull deer, plant trees, nurture saplings. In short, the people who will love them and care for them. As Gordon Gray Stephens said to me: 'You can't restore a rainforest without people.'

10

Bringing Back Britain's Rainforests

It was a bitingly cold winter's day, shortly before Christmas, when I drove to the edge of Bodmin Moor to meet a man named Merlin.

The setting could hardly have been more apt. Bodmin Moor in Cornwall is soaked in Arthurian legend, a landscape steeped in magic. To the east lies Dozmary Pool, a small lake whose peat-black waters are the rumoured resting place of Arthur's sword Excalibur. To the west lies the hamlet of Temple, its tiny chapel built by the Knights Templar, fabled guardians of the Holy Grail after the Crusades. And though the bleak expanse of Bodmin Moor boasts more Celtic crosses and standing stones than it does trees, the river valleys on its periphery give shelter to temperate rainforests.

The man I had come here to meet, Merlin Hanbury-Tenison, was the proud owner of one of them: the Atlantic oakwood of Cabilla. Pink-cheeked from the cold and wearing a flat cap and tan-coloured barn coat to keep out the chill wind, Merlin greeted me next to a cluster of farm buildings undergoing renovation. He owes his somewhat romantic name to his father, the explorer Robin Hanbury-Tenison – whose adventures in the Amazon rainforest led Robin to found Survival International, which campaigns for the rights of Indigenous peoples around the world.

But as we got talking, it quickly became clear that Merlin was no ethereal hippy, but driven and practically minded. An ex-army captain who undertook three tours of Afghanistan during his twenties, Merlin returned from service suffering from post-traumatic stress disorder. He took solace in nature at his family's 300-acre farm, Cabilla Manor, finding being outdoors in the woods and fields to be healing and restorative.

Yet as he and his wife Lizzie took over the running of the farm in 2018, he also found that all was not well with the land. Traditional farming on this marginal pasture was no longer making ends meet. Centuries of overgrazing had depleted the thin soils of nutrients. BSE and foot-and-mouth disease had caused Cabilla's cattle and pigs to be slaughtered in quick succession, and the distraught tenant farmer had attempted suicide. To turn things around, Merlin decided to chart an audacious new course. Rather than intensify production, squeezing the dregs from knackered soils, he opted instead to regenerate the land. And, in the process, he quickly acquired a new mission: to rebuild a rainforest.

At the heart of the farm is a hundred acres of ancient oakwood. Merlin took me to see his favourite view of it: a vertiginous cliff clad in woodrush and bilberry bushes, offering an awe-inspiring vista over the wooded valley below. Gnarled oaks protruded horizontally over the scarp, felted with mosses and playing host to pennywort and polypody ferns. Beard lichens dripped from their coiling branches. 'The oaks aren't single organisms, but entire cities,' marvelled Merlin. From the exuberance in his voice, I could see how being in the forest had an uplifting effect on him. His story reminded me of the Arthurian legend of the Fisher King: a wounded warrior, guardian of the Grail, whose sickness is bound up in the health of the surrounding land. When one heals, so does the other.

The rainforest soon began to work its magic over me, too. Beneath the oaken canopy at the bottom of the gorge rushed the Bedalder River. 'Bedalder is Cornish for "sweet water",' said Merlin. 'It really does taste sweet, it's so pure.' He told me of the wild swimming spots along it, and the trout that lurk in its depths. As we walked through the ancient woods, a raven croaked overhead; witches' butter, a black and gelatinous fungus, mushroomed from a dead bough; stinky *Sticta* lichens bubbled from the branches of willows. Merlin crouched down to peer inside a hollow tree, and gestured for me to look. The dark interior of the trunk glowed lustrous green. 'Goblin's gold,' he explained. 'A bioluminescent moss. *Schistostega pennata*. You find it down old mines.' But try bringing it out into the daylight, and the golden glow will disappear, like faerie treasure.[1]

The Domesday Book suggests there's been a wood here for a thousand years; peat core samples commissioned by Merlin point to one having been here for even longer. But great age brings with it senescence. The rainforest will die if it can't regenerate itself. So Merlin decided he wanted to turn 100 acres of rainforest into 300. Working with Lizzie and a team of ecologists, Merlin conjured up a plan for its restoration. This, he knew, wouldn't be something that could be achieved overnight. The Cabilla Cornwall website describes the undertaking as a 'thousand year project . . . to transform the Cabilla valley from an upland hill farm into a resilient temperate rainforest.'[2] Such long-term thinking may seem utopian, even grandiose. But given this fragment of rainforest has stood here for so long already, Merlin reasoned, oughtn't we be planning for its future another millennium hence?

The plan to regenerate Cabilla's rainforest starts small, with its acorns. The old oakwood still has mast years every five years or so, producing thousands of acorns – but, for decades, the seedlings that germinated were nibbled to the quick by deer and livestock. So an exclosure fence was installed to keep out browsing animals. Now, when fresh oak saplings sprout under the woodland canopy, Merlin and his team gently tie a small piece of yellow string around each of them, so they can be easily spotted. After a season or two they'll be ready for transplanting into surrounding fields. Merlin is also setting up a tree nursery, big enough to grow 40,000 trees at a time, gathered from local seed – including acorns from the 1,000-year-old Darley Oak, a venerable Methuselah that clings to life on the eastern edge of the moor. The rainforest looks set to expand through a mixture of planted saplings and natural regeneration.

But Merlin is also adamant that Cabilla will remain an active farm. The aim isn't simply to create closed-canopy woodland, but rather to create a series of glades in which woodland-edge species can thrive: wildflowers, insects, butterflies. 'We want to have a herd of twenty-five Highland cattle who'll be sent on this grazing circuit between the glades twice a year, alongside a dozen Exmoor ponies,' Merlin explained, showing me a sketch map of the planned layout.

Other animals, too, will ensure the regenerating rainforest is subjected to natural forms of disturbance. On our walk across the fields, we met Gloria the sow, who amicably nosed up against our welly boots, grunting and snouting in the churned-up mud. Gloria and her drove of Cornish black pigs are helping to break up the hard, compacted soils, creating conditions for other plants to germinate besides grasses. As a result, pearl-bordered fritillary butterflies are returning to the area, Merlin told me.

As we walked, I noticed the fields were also producing a fine crop of magic mushrooms: clusters of little liberty caps, poking their conical heads impishly through the turf. It reminded me of our adventures in Wales that past summer; of Jon Moses picking hallucinogenic fungi in Welsh rainforests, and reading about Gwydion, the shape-shifting wizard of Celtic mythology. 'Having a rich layer of mycelium in the soil is very important for its fertility,' said Merlin. 'We want to help restore it.'

Wistman's Wood on Dartmoor, with its stunted moss-clad oaks, is often thought of as being moribund. Yet in the past century, the wood has in fact doubled in size, thanks to natural regeneration arising from a temporary reduction in grazing pressures.

Lobaria virens, a startlingly green lichen with a shiny satin skin, loves the dampest parts of the Lake District, Wales, Scotland and the West Country. The orange discs are the lichen's fruiting bodies, called apothecia; they distribute spores, allowing for sexual reproduction.

Hart's-tongue fern *(Asplenium scolopendrium)* lolls from earthern banks, fleshy and glistening.

Beefsteak fungus (*Fistulina hepatica*) resembles, as its name suggests, a slab of raw meat – complete with dripping blood. Though not itself an indicator of rainforest, it thrives on oak trees, making the fungus a common sight in our Atlantic oakwoods.

The beautiful *Ptilium,* an uncommon moss of northern woods, glows gold and is shaped like an ostrich feather.

Ballachuan Wood on the Isle of Seil, western Scotland, is a prime example of vanishingly rare Atlantic hazelwood – and may have stood on this spot since the last Ice Age.

Though it resembles a bug-eyed monster straight out of science fiction movies, *Collema fasciculare,* or 'Octopus suckers', is in fact a jelly lichen found in the temperate rainforests of western Scotland.

One of the weirdest denizens of Britain's rainforests, hazel gloves fungus (*Hypocreopsis rhododendri*) is found only in rare Atlantic hazelwoods. It is thought to parasitise another fungal species, the equally bizarre glue crust fungus (*Hymenochaete corrugata*), which lives on decaying hazel branches.

William Wordsworth is often thought of as a poet of daffodils and sheep, but he was also arguably our first poet of the rainforests. He took inspiration from the magnificent Atlantic oakwoods of Borrowdale in the Lake District, shown here, and mourned how the region had been deforested over the centuries.

The oceanic liverwort *Bazzania trilobata,* which grows in large mounds in our Atlantic oakwoods. To me, with their segmented lobes and tiny 'legs', they resemble bright-green centipedes.

Liverworts thrive in our wettest woods. This species, *Scapania ornithopodioides,* has an unusual scent: to the bryologist Ben Averis, it smells 'like unwrapping a present'.

The twisted, octopus-like branches that characterise many of Britain's rainforest trees are on display in this photo of Black Tor Beare, one of Dartmoor's three surviving upland oakwoods.

Tree lungwort (*Lobaria pulmonaria*), a charismatic lichen of temperate rainforests, thought by medieval herbalists to be a cure for respiratory diseases. To my eyes, it looks like dragon scales, or the wings of the Jabberwock in Lewis Carroll's iconic poem.

A dying rainforest in Scotland, destroyed through overgrazing by deer and sheep, draws its last breath.

Indigenous activists from the Amazon – Sia Huni Kuin, Busa Huni Kuin and Narubia Werreria – perform a blessing ceremony in a Scottish rainforest, Cormonachan Community Woodlands in Argyll.

When Merlin's father started farming at Cabilla in the 1960s, the farm had employed eight people. But by the 2000s, it was providing work for just one person. Mechanisation has reduced the dependence on manual labour, whilst supermarkets have put the squeeze on dairy, meat and vegetable prices. Shifting diets have compounded the problem for hill farmers: Britons' meat consumption has fallen by 17 per cent in the past decade, and looks set to fall further as the public wakes up to the climate and health benefits of a more plant-based diet.[3] And then, of course, there's been Brexit. 'Upland farmers on Bodmin Moor are 80 to 90 per cent dependent on EU subsidies, but they're being phased out over the next few years,' Merlin reasoned. The current system pays farmers simply according to the area of land they farm; the bigger the landowner, the more taxpayers' money they tend to receive. But in future, farmers and land-owners will only get public money if they invest in public goods, like restoring habitats.

Merlin and Lizzie's decision to move away from traditional farming and develop a new model, therefore, has been driven as much by economics as anything else. 'We decided to flip the traditional farming calculation on its head,' said Merlin. 'Rather than woods and trees being seen as getting in the way of farming, we've made them central to our business.' Having experienced the benefits of being immersed in nature for his own mental health, Merlin saw an opportunity to diversify Cabilla's income streams by offering yoga retreats and forest bathing in the ancient oakwoods. For him, this isn't just 'woo-woo hippy talk', but underpinned by a growing body of science that proves what our hearts have long known: that contact with nature helps heal us.[4] Cabilla has a research programme working with Loughborough University to study the mental health benefits of being in nature.[5]

By focusing on producing less but better-quality meat, the farm now looks to tap into a growing demographic which sees meat more as a treat than a staple. And Merlin sees the UK government's reformed system of farm payments as a 'big opportunity to transition to a different way of doing things, including temperate rainforest restoration'. Now Cabilla employs nine local people, more than it did when Merlin's father first started out here – and Merlin hopes the numbers will grow still further.

Alongside the people returning to this landscape are some other, long-absent creatures. In 2020, Merlin reintroduced beavers to Cabilla, after their 400 year absence on these shores. The two beavers – a female nicknamed Sigourney Beaver, and a male known as Jean-Claude Van Dam – have recently had kits. Merlin took me to see the beaver enclosure in amongst a marshy thicket of willows. Whilst we didn't spot the beavers themselves, the gnawed tree trunks and huge dam made out of branches and twigs were clear indications that they'd begun to cast their spell over the valley.

Merlin's project to restore a rainforest left me feeling energised and inspired. And, though he remains a pioneer, he's not alone. A growing movement of people in Britain is becoming re-enchanted with our rainforest heritage. This final chapter is about how Britain can and must bring back its lost rainforests.

First, let me restate the case for *why* we should bring back Britain's rainforests.

A habitat that once flourished over perhaps a fifth of Britain has been reduced to scattered fragments covering less than 1 per cent of the country. The cause of this catastrophe is in no doubt: we cut them down – for fuel, timber and farmland, from Neolithic settlers through to twentieth-century foresters. Yet even though this process of deforestation took place over centuries, much of it long ago, it doesn't diminish the tragedy of their loss. As the legendary botanist E. O. Wilson, who died whilst I was writing this book, once said: 'Destroying rainforest for economic gain is like burning a Renaissance painting to cook a meal.'[6] He was speaking about tropical rainforest, but it applies equally to the lush and exquisite temperate rainforests of Britain. We have a moral obligation to try to repair the damage caused by our ancestors.

In doing so, we would find that temperate rainforests are amongst our best allies in other battles we face. To stave off the existential threat posed by the climate crisis, it's imperative that we not only cease burning fossil fuels but also restore the natural carbon sinks that we've destroyed. Rampaging wildfires and killer floods show that climate breakdown is already upon us: we've already pumped too much carbon into the atmosphere and need to start actively removing it. Tech entrepreneurs dream of building machines to capture carbon dioxide from the air, but they're overlooking the ones that nature spent millions of years designing: trees.

The trees of temperate rainforests, in common with those found in all woodlands, absorb carbon from the air and lock it up in their trunks and branches. The leafmould of the forest floor, too, accumulates carbon. And the extraordinary aerial ecosystems that distinguish temperate rainforests from their drier counterparts could also be playing a role in sequestering CO_2 from the atmosphere. The thick layers of moss and humus that form on the branches of Britain's rainforest trees likely act as carbon stores in their own right: generations of plants living and dying on every branch, generating a rich layer of soil carbon high in the canopy of the rainforest.

Urgent as the climate crisis is, even it pales before the parallel crisis we face of the sixth mass extinction. The biodiversity crisis gets far less press than climate change, but if anything it's bigger and scarier. It means nothing less than the unravelling of the web of life upon which we all depend. Britain is now one of the most nature-depleted countries on earth.[7] In the last fifty years, the UK has lost half of its farmland birds, seen a 40 per cent decline in woodland-edge butterflies and fallen prey to precipitous declines in insect populations.[8] Yet pockets of biodiversity still cling on. Though Britain's temperate rainforests may only be small, they support a disproportionate wealth of species. From tree lungwort to pine martens, pied flycatchers to blue ground beetles, ancient oaks to young rowans, Britain's rainforests play host to a dizzying variety of gorgeous life-forms.

Temperate rainforests are the very pinnacle of Britain's woodlands. They support around 500 species of lichen and over 160 species of mosses and liverworts.[9] Temperate rainforest is, as we have learned, rarer than tropical rainforest: the latter still comprises around 12 per cent of Earth's forest cover, whilst temperate rainforest accounts for just 1 or 2 per cent of it.[10] A report in 2018 by the Joint Nature Conservation Committee – an official government body – stated that Britain has 'international responsibility' for its temperate rainforest habitats. 'Woodlands along this Atlantic fringe are the European headquarters for lichens with oceanic and . . . hyperoceanic distributions,' it declared.[11] Britain's famously rainy climate, in fact, makes it the best place in Europe for this habitat to thrive.[12] When it comes to temperate rainforest, Britain truly is world-beating.

You also don't need to be a trained ecologist to find being in a temperate rainforest exhilarating. Britain's rainforests are simply awe-inspiring places to visit. Regularly spending time in nature is increasingly recognised as beneficial to our physical and mental health. Immersing myself in the dripping, viridian fastness of an Atlantic rainforest is the closest I've ever come to a spiritual experience. Small wonder that different cultures throughout history have venerated them as special places: the sacred groves of the Celts, the sense of the sublime celebrated by the Romantics. We need that awe and magic in our lives, too.

What's more, we can restore rainforest without impacting on food production. British rainforest appears to thrive on rocky screes, in steep-sided gorges and on thin upland soils: places where agriculture is already marginal, if not impossible. The government-commissioned National Food Strategy calculates that we could spare 20 per cent of England's least productive farmland for nature and only reduce food production by 3 per cent.[13] Much of that least productive land is in our western uplands, areas in which the climate favours rainforest. As Brits continue to reduce the amounts of meat and dairy they eat, it surely makes sense to return this surplus land to nature. Whilst the west of Britain is good at growing grass – as many farmers have long known – it is even better at growing rainforest.

Bringing back Britain's lost rainforests could also have repercussions far beyond these shores. The fight to save the Amazon rainforest has understandably galvanised the attention of environmentalists and the wider British public for decades. Today, with deforestation rates soaring under Brazil's far-right President Bolsonaro, fuelled by the west's taste for beef and poultry fattened on rainforest soy, the fate of the Amazon still hangs in the balance. The greatest hopes for the rainforest's future lie in Brazilians ditching Bolsonaro in favour of a new government committed to Indigenous rights, and for countries like Britain to stop propping up destructive agribusiness through our consumption of meat and dairy. But calls from British politicians for developing nations to halt deforestation would also carry greater weight if we hadn't so thoroughly deforested our own country.

Britain is, after all, guilty of double standards when it comes to nature conservation. We expect people in India and sub-Saharan Africa to coexist with apex predators like lions and tigers, but can't seem to stomach the idea of reintroducing wolves and lynx to our own countryside. We lecture countries in the Global South about ending trophy-hunting, but blithely tolerate huge sporting estates which continue to trash the uplands of Britain. We call on Brazil and Indonesia to save their rainforests, yet don't own up to the fact that Britain is a rainforest nation, too – just one that's already cut most of ours down. We might not immediately see this hypocrisy, but other countries do, and raise this in diplomatic negotiations.

In that respect, bringing back Britain's own rainforests is also a decolonial project – one which starts to hold us to the same standards that we expect of others. And for a wealthy, industrialised nation like Britain to show that its future development lies in *restoring* nature rather than destroying it – that could provide a powerful inspiration for a concerted global effort. When Boris Johnson called on world leaders at COP26 to 'protect and restore the world's forests', more might have heeded his entreaties had he added: 'and today Britain pledges to do its bit too, by protecting and restoring our lost rainforests'.[14]

The loss of our country's rainforests has been not only an ecological disaster but also a cultural tragedy. The fragments that remain are reminders of our lost heritage: an ecosystem that inspired some of our greatest myths and legends. They provide a link back to the deep history of this island; from the Wildwood encountered by the first peoples who settled here after the last Ice Age, down through stories told around campfires by generation after generation. The Atlantic rainforests are woven into Celtic mythology, the weird tales of *The Mabinogion*, the Arthurian romances, the body of literature that folklorists call 'the Matter of Britain'. In more recent times, they've inspired Wordsworth and Carrington, Conan Doyle and Tolkien, Romantic poetry and popular fantasy. They can inspire and rejuvenate us, just as they did our predecessors.

We Brits are notoriously obsessed by the weather and how much it rains here. So we may as well make a virtue of that fact, and restore the ecosystem that evolved specifically to exploit how rainy Britain is: our rainforests. Rainforests, though seemingly exotic, are also as British as a cup of tea.

So how do we do it?

As Merlin pointed out to me, restoring rainforests is a

multigenerational project, ultimately spanning centuries. It's much easier to destroy a habitat than it is to rebuild one. It takes time for a woodland as biodiverse as a temperate rainforest to develop: for rare communities of lichens and bryophytes to creep slowly along branches, finding just the right places in which they can thrive. But nature can also surprise us with its capacity to regenerate faster than we expect.

We need to start with what remains. The few precious fragments of rainforest left on these shores are like the lifeboats of a huge ship that sank beneath the waves. Yet these lifeboats now lie marooned, scattered to the four winds, with supplies running low. They need rescuing.

Some of our surviving rainforests are not even accorded legal protections. The extraordinary Millook Valley Woods in north Cornwall, for example, harbour what an eminent botanist calls an 'internationally important epiphytic lichen flora'. Yet the woods have never been designated as a Site of Special Scientific Interest, or SSSI, 'in spite of this international importance'.[15] Granting the woods this status would give them legal protection against future development and harm for all time. Fortunately, the rainforests at Millook are in the hands of the Woodland Trust, who look after them well. But other undesignated rainforest sites may not be so lucky. It's essential that the UK government and devolved administrations grant legal protection to all remaining rainforest fragments by designating them as SSSIs.

Yet many rainforest sites that are notionally protected on paper aren't always in good shape in reality. The temperate rainforest of Johnny Wood in the Lake District was recently found by Natural England to be in 'unfavourable – declining' condition as a result of overgrazing by sheep.[16] In Scotland, up to a million deer menace the nation's rainforests, their numbers kept unnaturally high by the country's sporting estates and an absence of natural predators. Invasive rhododendron has now spread over a quarter of a million acres of Britain, threatening to suffocate the rainforests in which it takes root.

We need to grapple with all these threats simultaneously in order to give our remaining rainforests a chance. Deer numbers in Scotland and across Britain need to be controlled through culling and, ideally, the reintroduction of natural predators. There must be a concerted country-wide campaign to eradicate rhododendron. And we have to wrestle with the problem of rainforest mutton – Britain's equivalent of the rainforest beef that's been destroying the Amazon. Sheep numbers in Britain's uplands may well decline anyway as subsidies shift so that sheep farming becomes even less economic than it is already, and as we continue to eat less and less meat. But, in the meantime, we've got to prevent sheep from nibbling away the remnants of our rainforests.

Some of our rainforests now lie dormant beneath a blanket of conifers, existing only as seed banks buried in the forest floor. Unforgivably felled by twentieth-century foresters who sowed pine plantations in their stead, they wait patiently to be resuscitated, like Sleeping Beauty. The national mission to restore Plantations on Ancient Woodland Sites has been championed by the Woodland Trust for decades, but it is agonisingly slow work. And though modern forestry has thankfully stopped felling ancient woods, there are still plenty of landowners unwilling to restore their profitable plantations to flourishing rainforests. It's hard to completely kill a rainforest: phoenix trees rise from the ashes, seed banks spring back into life. Yet wait too long to restore a felled woodland carpeted with conifers and even its buried seeds will give up the ghost.

The biggest threat to our rainforests' long-term survival is that they remain fragmented, isolated islands. Ancient relics like Young Wood in the Lakes and Wistman's Wood on Dartmoor cling on, but they lie marooned in green deserts, far from help. This reduces their resilience to pressures like overgrazing, and severs them from natural processes like seed dispersal. Few jays and squirrels make it to these remote outposts of rainforest to help disseminate acorns, for example. Fragmentation also limits the ability of rainforest species like lichens to spread. The reproductive spores and soredia of some lichens like tree lungwort can't travel very far, even relying on raindrops to allow them to drip from branch to branch. This reproductive strategy hits a brick wall when you run out of rainforest. As a result, our rainforests play host to dwindling populations of rare species that are increasingly vulnerable to freak events like storms.

The ecologist George Peterken, a true sage of the woods, has written about the effects of ecological isolation on ancient woodland. 'This fragmentation has had profound effects on the native fauna and flora,' he explains. 'Less mobile species have been broken up into a scatter of separate sub-populations, between which there is little exchange of individuals.' Think of the disjunct populations of Lobarion lichens and filmy ferns which cling on in remote gorges and ghylls, from Cornwall to Sussex. The answer, Peterken argues, is to reconnect our woods: 'New woodland is needed to expand existing woods and fill the gaps between them, thereby allowing more movement of woodland species across the landscape.'[17]

It's worth reiterating at this point that whilst tree-planting has become something of a civic religion in modern Britain, forests have, as we have learned, regenerated naturally for millions of years. Jays, squirrels and other animals are nature's tree-planters, and other trees self-seed through wind dispersal: an ash will produce hundreds of twirling ash keys. If we want to restore our rainforests, natural regeneration like this is by far the best option. Planting trees still has a role to play where there's no nearby seed source. But planting, even when done sensitively with broadleaved species, tends to result in a woodland where all the trees are of the same age, with a limited mix of species, and with saplings planted in neatly spaced rows. This bears little resemblance to a wild wood, which will have a far richer mix of species of trees, shrubs and understorey plants, as well as a better 'age structure' ranging from young saplings to veteran giants of the forest.

Rainforest restoration, therefore, is best done by starting with the fragments we have left, and letting them spread. Begin with the ancient woods, burgeoning with their wealth of species, and let them expand into the surrounding landscape. A self-willed rainforest advances in waves, like the animated trees in the old Welsh poem 'Battle of the Trees', or Tolkien's march of the Ents: thorny scrub in the vanguard, followed by birch and rowan, with ash and oak lumbering behind. This is no mere fantasy: I have watched rainforests on manoeuvres, their slow but inexorable movements up hillsides and along valleys becoming apparent when you flick between old maps and modern aerial photographs. Places like Lustleigh Cleave on Dartmoor's eastern flank have gone from bare common to lush rainforest in a matter of decades.

But we still need to intervene to give rainforests the space to breathe. Natural regeneration in Britain is continually thwarted by the teeth of grazing sheep and browsing deer, who exist in unnaturally high numbers. So to let our rainforest fragments regenerate, we need to control grazing and browsing on their edges. One method is to install exclosure fencing around the perimeters of our rainforests, leaving a wide 'buffer zone' into which scrub and saplings can spread. Another way is to fit livestock herds with Nofence technology, managing their grazing locations with GPS-guided collars. Of course, such technology will be of no use for controlling wild deer, who will also leap all but the highest fence. So deer culling must be increased, at least until we have the courage and public consensus to reintroduce more natural predators like lynx and wolves. Together, these methods can give our rainforests the space to regenerate.

A recent study has shown that when grazing is controlled,

ancient oakwoods can naturally regenerate out to distances of 100 to 150 metres in just twenty-three years.[18] From the mapping work I've done with GIS expert Tim Richards (see Chapter 4), we can see there's at most around 333,000 acres of temperate rainforest left in Britain today. Tim and I wondered how much more rainforest we might have if landowners and farmers agreed to prevent overgrazing around their edges, and let natural regeneration take its course. So we had a go at mapping this – taking our map of existing rainforests, and drawing in 'buffer zones' around them: a 'restoration halo', if you like. We also applied some sensible constraints, by screening out peat soils, species-rich grassland and high-quality farmland.

The results were startling. If we allowed those rainforest fragments to regenerate and spread outwards by just 150 metres on all sides, we could almost *double* the area of temperate rainforest in England alone – without impacting on other precious habitats or productive farmland. We also think this figure would hold true for Britain overall.[19] Doubling Britain's rainforest cover in less than a generation – now *there's* a goal to aim for.

Many conservationists are understandably concerned that a rush to create new woodlands might end up damaging other precious habitats. Ensuring we get the 'right tree in the right place' has become an oft-repeated mantra in environmental circles, and for good reason. In the 1980s, huge areas of peat bog in the Flow Country – a broad expanse of deep peat in the far north of Scotland – were drained and ploughed to plant Sitka spruce: an ecological disaster which likely resulted in far higher carbon emissions being released from the damaged peat than any CO_2 sequestered by the growing trees. Trees tend not to naturally colonise wet peat, which is why the foresters had to drain the bogs first to make the conifer plantations grow. Where trees and bushes do grow on deep peat, they tend to be patchy, rather than closed-canopy woodland. Despite our knowledge of this, we continue to see gross mistakes being made to this day; in 2020, a hundred acres of peat bog and wildflower meadow were ploughed up by a Cumbrian landowner wanting to plant trees.[20]

So how might we avoid such errors when restoring rainforest? As a minimum, any woodland creation project should begin with a thorough ecological survey. Thanks to years of government austerity, local authorities and green watchdogs like Natural England no longer employ sufficient trained staff to screen these sorts of projects, and so need better resourcing. Another way to reduce the risk is simply to rely more on natural regeneration, rather than instantly reaching for spades and saplings.

We can also target efforts to restore rainforest by learning from the landscape. Take a look at the shape of some ancient woods on an Ordnance Survey map. Sometimes the field boundaries are curved, like the shape left in a sandwich after you've taken a bite out of it. This can be the mark of historical deforestation, a process of woodland clearance called 'assarting'. It doesn't necessarily preclude other valuable habitats having developed on the woodland edge in the centuries since it was deforested, but it can at least point to where trees would once have been. Other clues may lie in the old place names that refer to 'ghost woods' now long since gone from the landscape, many of which I've featured in this book – from 'Beares' on Dartmoor and Birker Fell in the Lakes to Brian Choille in the Highlands.

Another way to guide efforts to bring back rainforest is to look at where bracken grows. Bracken is often disliked by farmers and conservationists alike: when it appears in an upland pasture, it quickly overwhelms it, replacing grassland and heath with a monoculture that shades out other plants. But rather than bracken being a menace, I think it could be telling us something important. As we learned in Chapter 2, Ian Rotherham, the ecologist who's pioneered the study of 'ghost woods', considers bracken to be an indicator of former woodland soils. 'Under bracken there's gold,' goes the old hill farm saying.

When I met with Lee Schofield, site manager at RSPB Haweswater in the Lake District, he suggested I track down some bracken maps to help guide rainforest restoration efforts. Since meeting with Lee, I've been sent bracken maps for both Dartmoor and the Lakes. They reveal that bracken now covers at least 35,000 acres of the Lake District, or 6 per cent of the National Park, and a similar percentage of Dartmoor National Park. What's more, many of these bracken stands extend outwards from old-growth temperate rainforests – forming haloes around Wistman's Wood, Black Tor Beare and Piles Copse on Dartmoor, and drawing a huge arc around the Atlantic oakwoods of Borrowdale in the Lakes. The presence of this bracken likely points to the former extent of these woodlands, feathering up hillsides and extending along valleys. Vast bracken monocultures are not productive farmland, and encouraging trees to grow on them carries much lower risks for biodiversity and carbon than doing so on peat or species-rich grassland. So why not let our temperate rainforests expand into the surrounding bracken?[21]

When seeking to restore rainforest, our aim mustn't simply be to create wall-to-wall closed-canopy woodland. Instead, our ambition should be to generate a rich mosaic of habitats, combining rainforest with wood pasture, wetlands and wildflower meadows. Dense woodland should be studded with glades that let in plenty of light. And whilst an overgrazed rainforest is an unnatural habitat, a woodland bereft of all animals is equally unnatural. We need disturbance in our rainforests for them to be healthy, dynamic ecosystems. Pulses of conservation grazing by native breeds of cattle help let in the sunlight needed by rare lichens. Other natural processes currently absent from our rainforests could be reinstated by reintroducing missing species. Wild boar could trample down bracken stands, and as they rootle about in the forest floor they could provide useful 'ground preparation' for saplings and plants. Pine martens could help control grey squirrel numbers, lessening the damage they can do by ring-barking trees; and lynx and wolves could help reduce overbrowsing by deer.

Restoring Britain's rainforests on a large scale hasn't been done before. We certainly don't have all the answers yet, and we'll need to learn as we go. But we know broadly where to start. What's been missing thus far has been the will to do so – from landowners, politicians and the public. Happily, however, that's now starting to change.

I managed to turn up twenty minutes late to my first meeting about restoring Dartmoor's rainforests. I'd taken a wrong turn after exiting the M5, and got lost en route to the meeting venue: the National Trust's Killerton Estate, just north of Exeter. Crowds of tourists were starting to arrive as I hurried through the visitor entrance and across tree-studded parkland towards the Trust's offices in a converted wing of the imposing Killerton House. The estate had been given to the National Trust in the 1940s by baronet-turned-socialist Sir Richard Acland, along with Horner Wood, a majestic swathe of Atlantic oakwood on Somerset's north coast.[22] It seemed an appropriate place for raising the subject of rainforest restoration.

After darting through the entrance and navigating the maze of opulent corridors in the eighteenth-century mansion, I found the meeting room at last. With discussions already in full swing, I murmured my apologies and took a seat at the polished oak table. I'd heard about the meeting on the grapevine – it was billed as a discussion about a possible nature recovery project on Dartmoor – and had pleaded for a last-minute invitation. I'd detected some initial reticence about letting a tweeting troublemaker like me join a private gathering of landowning bodies, but the organisers had kindly relented. Now, surrounded by veteran conservationists affiliated to big organisations, I started to fret about what I could usefully add to the discussion. A fresh round of introductions for my benefit revealed the assembled party included many of Dartmoor's largest landowning bodies, from Dartmoor National Park Authority to the Devon Wildlife Trust and the Woodland Trust, alongside our hosts the National Trust. 'Campaigner and author, here in a personal capacity,' I said when the round of introductions reached me. 'No land to offer, I'm afraid.'

But though I had no land, it transpired that there might be other things I could bring to the table. The meeting moved on to what the assembled organisations wanted to achieve collaboratively. Conversations veered from reconnecting river catchments to restoring peat bogs; from protecting Rhôs pasture to boosting bird populations. All of it sounded very worthwhile, but it seemed to lack a common thread, a cause that might inspire public support and encourage other landowners to get involved. Sensing a lull in the conversation, I decided to chip in. 'What about focusing the project on restoring Dartmoor's lost rainforests?' I suggested.

There was a pause; I glanced with trepidation around the table, seeing some raised eyebrows and quizzical expressions. But as I pressed on, outlining what I saw to be the benefits of reconnecting rainforest fragments on Dartmoor's eastern edge, I could see others had already been thinking along similar lines. Harry Barton, head of the Devon Wildlife Trust, was nodding enthusiastically. So too were Alex Raeder, southwest lead for the National Trust, and Eleanor Lewis from the Woodland Trust, whom I'd met previously to look around Fingle Woods. After all, I reasoned, the organisations present at the meeting collectively owned or managed most of Dartmoor's remaining temperate rainforests – from the Atlantic oakwoods on the River Teign, and the ancient woodlands along the Bovey near Lustleigh Cleave, to the rainforests that clothe the Dart valley. Why not seek to reconnect these sites, creating a 'rainforest belt' down Dartmoor's east flank?

Not everyone present was so keen. I remember one attendee sounding distinctly dubious about using the term 'rainforest' in the same breath as Dartmoor, the scare quotes audible in their response. But as Eleanor pointed out, the term has the dual merit of being ecologically accurate whilst also having broad public appeal. I do sometimes wonder whether conservationists hamper their own cause by being needlessly exclusive and technical in their language. It's all very well talking about 'restoring W17 woodland', but you might struggle to get much public buy-in. The meeting wrapped up with attendees taking away a number of action points. To my pleasant surprise, I was invited back to the next meeting.

As one of the actions, Harry had offered to write a 'vision statement' for what the group wanted to achieve. I was excited to see that rainforest restoration was at its heart. 'We want to create a landscape that shows England's uplands at their very best and sets an example of how they can be restored across the country,' it read. 'The stand-out habitat will be temperate rainforest.' The project, it was proposed, should involve 'allowing temperate rainforest, wet woodland and scrub to creep up hillsides'. This was all music to my ears.

For my part, I'd volunteered to map who owned the land across Dartmoor's eastern edge.[23] The conservation bodies who'd initiated the meetings owned around 2,500 acres of it in an almost contiguous single block, encompassing Yarner Wood, land around Haytor, and the Devon Wildlife Trust reserve at Emsworthy. They were joined by the privately owned Leighon Estate, whose conservation-minded owner added another 500 acres. This core area was to be the focus of ecological restoration efforts. But I was keen to encourage the group to also be more ambitious, and reach out to other landowners and farmers across a wider area. There was agreement that, in future, the project should seek to reconnect the temperate rainforests of the Bovey in east Dartmoor with those along the Dart, some ten kilometres to the south. I knew from the bracken maps I'd been sent of Dartmoor, and from exploring the area on foot, that much of this intervening land was already scrubbing over with bracken, hawthorn and rowan. If we could just get a few more landowners around here willing to restrict grazing, we'd have a belt of regenerating rainforest stretching for miles.

Sitting in on the meetings brought home to me some of the challenges of trying to restore habitats over entire landscapes. To do serious ecosystem restoration, you simply have to work at a very large scale. Many natural processes play out over large areas: rivers need space to wiggle, trees disperse their seeds across hundreds of metres, predators roam over huge distances in search of prey. Fitting this into a human-dominated landscape almost inevitably means securing collaboration from multiple landowners and farmers, even in a country so dominated by large landed estates as England. And getting agreement between landowners, even ones that broadly agree on conservation objectives, takes time and can be slow and bureaucratic. That's doubly the case when it comes to seeking funding for such projects. At the time of writing, the group is in the process of applying for a government grant under the new Landscape Recovery scheme being piloted by the Department for the Environment, Food and Rural Affairs (DEFRA). I'm keeping my fingers crossed.

As these meetings were taking place, I was also approaching other farmers and landowners around England, trying to spark enthusiasm for rainforest restoration. Dartmoor's biggest landowner is the Duchy of Cornwall, the ancient estate owned by the Prince of Wales. Its landholdings encompass both Wistman's Wood and Black Tor Beare, two of the three remaining upland oakwoods on Dartmoor. When I'd first started researching Britain's rainforests, I'd approached the Duchy to ask if it had any plans to allow Wistman's Wood to regenerate and spread. I'd come away from my first call feeling excited to hear that it was at least interested in allowing the wood to expand in future.

But as the months passed, with successive calls and correspondence seeming to result in nothing concrete, I became disappointed by the glacial pace of change.

As time ticked on, I was cheered by the activist group Wild Card's brilliant campaign calling on the Royal Family to rewild their estates. A letter organised by the group – and signed by BBC presenters Chris Packham and Kate Humble, as well as a former Archbishop of Canterbury, amongst others – pointed out that 'large parts of Dartmoor, owned by the Duchy of Cornwall . . . would, if nature took its course, be covered by rare temperate rainforest'. Surely, the signatories urged, Prince Charles's longstanding commitment to the environment would lead him to change the way his estate was run. 'We are a rainforest nation bereft of its rainforests – and in desperate need of leadership that shows how it can be restored,' they wrote.[24]

If the Duchy was hesitating to act, out of concern for what its tenant farmers might think, the conversations I had with Duchy tenants suggested it had little to fear. I met with Naomi Oakley, who farms one of the Duchy's tenancies on the eastern side of the moor. On my arrival, I was greeted by Naomi's four excitable dogs and, more unusually, a friendly Icelandic sheep called Tintin, who seemed to think he was one of the pack and followed us around for the whole visit. (Clearly, not *all* sheep are bad.) Naomi showed me the wildflower meadow she and her husband Mark had created, bursting with the yellow blooms of bird's-foot trefoil, purple spikes of bugle, and lesser butterfly orchids. Their farm has to be one of the best on Dartmoor for nature, and Naomi and Mark are keen to do far more: they're letting willow scrub regenerate along streams, have re-wetted fields too marshy to graze, and would happily reintroduce beavers if their landlord lets them.

What's more, Naomi and Mark are part of a group of Dartmoor farmers also thinking on a landscape scale about rainforest restoration. They are members of a 'farmer cluster': a voluntary grouping of Duchy tenants and owner-occupiers keen to restore habitats and wildlife over a large swathe of eastern Dartmoor.[25] Most intriguingly of all, I was told some farmers were starting to talk about creating a 'Wistman's II' on the banks of the East Dart River, a younger sister to the legendary wood on the West Dart. A second Wistman's Wood! Was it possible? Could it be done? If the Duchy ends up agreeing with its tenants, then I can't see why not. There are certainly the right conditions on the ground: the land is in a newtake, boxed in by walls that could help with controlling grazing. And a swathe of bracken carpets the valley of the East Dart above Postbridge, a tantalising hint of a rainforest that wants to live again.

Although creating a second Wistman's Wood seems an audacious idea, I was astonished to discover that someone had already suggested doing it a generation ago. Whilst researching this chapter, I came across an old report published in 1997 by English Nature, the predecessor body to Natural England. The paper, entitled 'Developing New Native Woodland in the English Uplands', sought to map suitable locations for regenerating broadleaved woods on Dartmoor. In an uncanny foreshadowing of the conversations I'd been having, the report's authors suggested that 'most realistic opportunities for substantial woodland expansion will . . . arise in moorland edge locations', where existing woods afford 'good potential for natural regeneration'. In such places, the report argued, 'reduction or removal of grazing animals would be all that would be required to extend existing woodlands'. Most extraordinary of all was the authors' assessment there was potential 'perhaps even for the creation of something which might one day come to resemble the highly valued, high altitude Wistman's Wood'.[26]

Reading these words at first gave me hope, and then left me with a sense of dread. Was it true that history just moves in circles, with the same ideas going round and round without anything ever really changing?

Except that things *have* changed. What was a researcher's mere fantasy twenty-five years ago is now being seriously discussed by Dartmoor farmers, whose predecessors would never have dreamed of entertaining such ideas. The Duchy of Cornwall, the estate owned by the heir to the throne, is talking about letting Wistman's Wood regenerate and spread. A swathe of landowners on Dartmoor's eastern flank are mobilising to restore temperate rainforest. Dartmoor National Park Authority's new plan for the park includes a commitment to creating 5,000 acres of 'new valley native broadleaf woodland'.[27] 'There is a growing realisation that many areas of the UK, including Dartmoor, would naturally be more wooded,' declares a local Devon charity called Moor Trees.[28] And although the pace is still too slow, the growing appetite for change is palpable.

In other parts of Britain, too, there is a growing movement to bring back our rainforests. Scotland – invariably ahead of England and Wales on most things – is at the forefront, with a highly organised coalition of groups, the Alliance for Scotland's Rainforest, already making demands of the Scottish government.[29] In Wales, the Celtic Rainforests project – spearheaded by the Welsh Woodland Trust, Coed Cadw – is leading efforts to eradicate rhododendron and to practise conservation grazing in rainforests.[30] In the Lake District, Lee Schofield and the RSPB continue their work to restore temperate rainforest at Haweswater, whilst the National Trust have plans to create greater connectivity between the Atlantic oakwoods they own in Borrowdale. And whilst all of these nations and regions have their own conflicts – between landlords and tenants, farmers and environmentalists, commoners and rewilders – there's also, I hope, growing consensus.

We can also draw hope from burgeoning international momentum to restore lost rainforests. In Ireland, the farmer and writer Eoghan Daltun has inspired thousands with the lush temperate rainforest he looks after in West Cork. After installing an exclosure fence around the surviving ancient woodland in 2010, Eoghan says that the surrounding land experienced an 'explosion of luxuriant growth and biodiversity', with the rapid natural regeneration of trees.[31] In New Zealand, the botanist Hugh Wilson has spent thirty years restoring 3,000 acres of temperate rainforest on the Banks Peninsula. The wiry, white-bearded Wilson became the unlikely star of a viral documentary in 2019, *Fools and Dreamers*, which told the story of his quest to allow scrub and trees to reclaim an area of knackered farm-land. Once again, removing grazing livestock was key. At first, the project received short shrift from neighbouring farmers. 'They basically thought we were naive greenies from the city,' Wilson recalls. But three decades on, 'I don't think there's a single farmer who's not backing [us] now.'[32]

Inspiring as these wonderful projects are, we can't just rely on a few enlightened landowners and maverick botanists to single-handedly restore our lost rainforests. We need political change, too. That's why in 2021, in parallel with writing this book, I started a campaign calling on the UK government to bring back Britain's rainforests.

Before 2021, political awareness of Britain's rainforests at Westminster had been negligible. No MP had ever mentioned them in Parliament, and there wasn't a single government-issued document that referred to them. Environmental NGOs with interests in temperate rainforests had focused their efforts on practical conservation work and ecological surveying rather than on political campaigning.

It was the activist Joel Scott Halkes, co-founder of Wild Card, who first suggested to me the slogan 'bring back Britain's rain-forests', planting in my head the seeds of a campaign. Later, I was approached by Isabella Gornall, founder of the environ-mental consultancy Seahorse, who offered to work with me to lobby government. Together with her team, we embarked on a frenzy of activity: meeting with MPs, speaking with numerous civil servants, writing blogs, starting a petition on the Parliament website, and sending briefings to ministers. Our key asks were simple: get ministers to declare they wanted to restore Britain's lost rainforests, and push the government to draw up a 'Great British Rainforests Strategy'.

Underneath those headline demands have been a series of more detailed policy asks. First, it's clear that far too little has been done to properly protect our precious remaining remnants of temperate rainforest. As the ecologist Dominick DellaSala notes, Britain is almost unique globally in 'having no strictly protected rainforests'.[33] As we've seen, even Britain's most highly protected rainforests – those designated as Sites of Special Scientific Interest – are often in bad shape, suffering from the effects of overgrazing and invasive species. Plenty of other rainforest sites lack even these paper protections. It's absolutely imperative that the UK government gives its environmental watchdog, Natural England, the direction and resources to designate all remaining rainforests in England as SSSIs. The Scottish and Welsh governments should do likewise with their respective bodies.

This enhanced protection must go hand-in-hand with a Britain-wide drive to eradicate rhododendron from our rainforests. Too many efforts to date have been piecemeal and under-resourced. Since rhododendron is a zombie weed – it keeps coming back from the dead – longer-term government funding is needed to make sure it stays dead. And given that Britain's rhodie infestation was foisted upon us by a small number of huge sporting estates, it should be made a legal duty for large landowners in the rainforest zone to remove rhododendron from their land.

But simply protecting the rainforests we've got left is a recipe for continued decline. We need to go further and begin the long project of rainforest restoration. Government farm subsidies have to date funded rainforest destruction, by encouraging the overgrazing of marginal land in the Lake District, on Dartmoor and elsewhere. As it completes its post-Brexit reforms of farm payments, the UK government should make funding available for rainforest restoration under its new Environmental Land Management schemes (or ELMs). These new payments are intended to reward farmers and land managers for *restoring* nature, rather than diminishing it. The top tier of ELMs – called 'Landscape Recovery' – holds the most promise for this, because it's focused on the kind of large-scale ecological changes that will be needed to reconnect rainforest fragments. A set of pilots to trial the new Landscape Recovery programme has recently begun: it would be fantastic if the government funded a rainforest restoration project as one of these pilots.

The Forestry Commission has also recently opened a new grant, the English Woodland Creation Offer, which for the first time funds the natural regeneration of woodland. But it contains a critical flaw: it artificially limits funding for natural regeneration to land within just seventy-five metres of a seed source. As we've seen, studies show that jays and squirrels help disperse acorns and other tree seeds over far greater distances. Doubling this threshold to 150 metres, in line with the science, would encourage farmers and landowners to set aside considerably more land on the edges of rainforests, giving them the space to breathe and spread. Farmers would then have a clear incentive to control grazing, either through exclosure fencing – whilst maintaining public access to land by putting in gates – or by fitting GPS collars on their livestock.

A host of other policies should also be considered to help restore Britain's rainforests. Councils need to include temperate rainforests in their forthcoming Local Nature Recovery Strategies, whilst National Park Authorities ought to build them into future management plans – particularly for western National Parks like Dartmoor, Exmoor, the Lake District and

Snowdonia. Commons, as we've learned in earlier chapters, present their own particular challenges for habitat restoration – not least the difficulty of getting commoners to agree to restrict grazing and allow the commons to regenerate. Where it's possible for commoners to agree to do this, the process could be made less bureaucratic: currently, putting in livestock exclosures covering more than 1 per cent of a common requires approval from none other than the Secretary of State. It seems a little excessive to be involving a minister of the Crown in such affairs: the percentage threshold should surely be raised. And to allow more people to get involved in restoring rainforests, we should be legislating for a Community Right to Buy in England and Wales which matches the powers of Scotland's community buyout laws.

In the longer run, public education is also key: part of the problem is that we're simply unaware Britain possesses rainforest. Cut off from nature, our culture suffers from 'shifting baseline syndrome'. We're so accustomed to ecological loss as to have forgotten what we once had. Yet we can help cure this environmental myopia – as well as our broader blindness towards plants – by more people taking up botany as a hobby, and by teaching natural history in schools, as it used to be.

It's early days yet, but our campaign has already secured some early wins. In late 2021, as a result of our campaigning, the government announced it would set up a new £30 million 'Big Nature Impact Fund' to 'support and expand England's temperate rainforests', the first time it had ever even mentioned this precious habitat.[34] A subsequent article in the *Telegraph* quoted a DEFRA spokesperson as saying: 'Temperate rainforests are globally important and highly biodiverse habitats. DEFRA is committed to expanding and protecting them.'[35] Not long afterwards, Anthony Mangnall, the Conservative MP for my home town of Totnes, became the first Member of Parliament to raise the subject of Britain's temperate rainforests in Parliament.

In response to his parliamentary questions, Environment Minister Rebecca Pow acknowledged the 'international importance of temperate rainforests', noting that they 'support rich assemblages of species' and pledging to do more to fund their restoration.[36] I'll certainly be holding the government to its word on this.

Yet this is only just the start. Clearly far more is needed, from politicians and from landowners. And that's where you come in. To win, we need many more people to speak out and take up the cause. A generation ago, the rallying cry of 'Save the Rainforest' became an environmental cause célèbre. Today, we must go one better. Rainforests aren't just someone else's problem; they're right here on our doorsteps. And we shouldn't simply seek to save them, but to restore them.

Our mission must be to bring back the lost rainforests of Britain.

11

Ghost Hunting

It was a fine spring day up on Dartmoor, and I was on a hunt for ghosts.

More precisely, I'd come to look for the ghostly remains of a long-lost wood. In his 1909 *Guide to Dartmoor*, William Crossing speaks of a wood that once existed on the flanks of Fur Tor, one of the most remote places on the moor. 'Below it, on the Cut Combe side, is a spot known to the moormen as Fur Tor Wood,' he wrote. 'The name seems to point to the former existence of trees in this sheltered hollow, and the discovery a few years ago of oak buried in the peat near Little Kneeset proves that they once grew around here.'[1]

This tantalising reference had nagged at the back of my head over the long winter months, tugging at my curiosity. It'd been a long while since I'd ventured that far out on to the moor. The last time had probably been when doing the Ten Tors hike as a teenager, and all I remembered of it was a vast, barren expanse of grass stretching as far as the eye could see. The idea that it once supported another Atlantic oakwood – another Wistman's Wood or Black Tor Beare – gnawed at my imagination.

Then, as the evenings grew longer and the forecasts became less rainy, my itchy feet were getting unbearable. It was time for an adventure. I met my friend Lewis in a Totnes pub and with pints of beer in hand, we pored over maps of Dartmoor, plotting our best route. Fur Tor lies almost exactly in the centre of the northern moor, about eight kilometres from the nearest road; more like ten when factoring in rough terrain. But with lighter evenings and good weather forecast for that weekend, we reckoned we could do the round trip in a day, saving us the need to pack tents. And, because we now had an additional mission in mind, we decided to pay a visit to Wistman's Wood along the way.

So, armed with backpacks weighed down only with a vast array of snacks, Lewis and I laced up our boots and set off from the car park at Two Bridges. The unseasonal sunshine had burned away the mist that often wreathes Wistman's Wood, but not its mystery: I still feel something akin to an electrical charge every time I get close to it. Ducking momentarily beneath the low canopy of the oaks with their curtains of lichen, we squatted on our haunches, scanning the leafmould that had accumulated in amidst the boulder scree. 'Have you lost something?' asked a passer-by. 'No, it's okay thanks,' I replied. 'We're looking for acorns.'

'Found one!' called Lewis, from behind a mossy oak. 'Look, this one's even started to germinate.' Eventually, we each gathered a small handful of acorns, each one a riposte to the idea that Wistman's Wood is moribund and infertile. But the surrounding wasteland, bereft of saplings but abounding in sheep, pointed to the deeper sickness afflicting the wood. If left where they had fallen, not one of these seeds would develop into adult trees.

Carefully storing the acorns in our rucksacks, we set off again, clambering up Longaford Tor above the wood to get a view of our route. The tor comprised an impressive collection of precariously balanced rocks, ziggurats constructed by the wind and rain. As we paused to drink coffee from flasks, Lewis, a geographer by training, told me about the geology of Dartmoor.

'It's a huge extrusion of granite, called a batholith,' he said. 'About 300 million years old. During the last Ice Age, a lot of this rock was exposed to intense weathering, freezing and thawing over and over. That's what's created the weird shapes of the tors. It also sent scree tumbling down the hillside, to create the boulder field that Wistman's Wood now grows in.' As I listened to Lewis, I looked down at the rock I was sitting on, its crystals of quartz glistening in the sun. But the surface of the boulder was mostly made up of living things: a crust of lichens and mosses, like a flattened coral reef, some white, some radioactive yellow.

Away from the sheltering ecosystem of the tor, a cold breeze blew unimpeded across the moorland desert that stretched before us. Fur Tor lay far ahead, beyond the horizon. In between us and it lay mile after mile of sere grass, blown bone-dry and skeleton-white by the endless wind. Tussocks of *Molinia*, purple moor-grass, conspired to twist our ankles and made us stumble. Part of me loves the barren nature of Dartmoor: its wide-open skies, its remoteness from civilisation. But in all our hours of walking, we saw only a pair of ravens and a single, startled snipe. No jays or squirrels would bother burying acorns out this far. 'There's a bleak beauty to it,' Lewis agreed when we stopped for a sandwich, hiding from the wind in the lee of a hill. 'But some people have romanticised the bleakness. They seem to love how depressing it is. A bit like listening to Morrissey.'

In such a featureless landscape, we were grateful for any aid to navigation, even man-made ones like the MOD's red-and-white range markers. The wind soughed through the never-ending hummocks of moor-grass. As we trudged on, it occurred to me that we were not on a moor at all, but wading through a shallow sea, the grasses rippling like cresting waves. Presumably the same thought had struck previous explorers of the moor, because our route took us through an old peat-cutting with the decidedly maritime name of the North West Passage. THIS STONE MARKS A CROSSING THROUGH THE PEAT, intoned a small granite marker, embedded between eroding peat hags.

At last, Fur Tor hove into view ahead of us. I had half hoped to see a cluster of trees still clinging on, some sign that this landscape still supported rainforests. There was nothing of the sort, of course: only a vast expanse of grass, the sward clipped low by a flock of sheep.

But there was a clatter of boulders, cascading down the near side of Fur Tor towards the Cut Combe Water. *What are the roots that clutch, what branches grow / Out of this stony rubbish?*[2] None today; but once upon a time, this rocky scree had perhaps given shelter to oak and rowan, birch and ash, before they'd succumbed to the axes of tin-miners and the depredations of livestock. Reinvigorated by the sight of our destination, Lewis and I quickened our pace, scrambling over tufts of sedge.

The boulder field below Fur Tor resembled the collapsed ruins of some long-gone civilisation: oblongs and triangles of stone, tumbling over one another. I found, sandwiched between two megaliths, the delicate fronds of filmy fern – an oceanic species, more usually found in a rainforest than in the middle of a moor. Here and there, the stones had fissured, creating crevasses in which a small layer of soil had gathered. Into this thin loam we pressed the acorns we'd collected in Wistman's Wood. It was, we knew, a futile gesture: even if the acorns were to germinate, they'd be eaten at first bud by the sheep that frequented the tor. But perhaps the boulders would provide a protective carapace for a while, as they had for Dartmoor's other oakwoods. Perhaps Fur Tor Wood might yet live again.

Turning to leave, we spotted something gleaming white, half-buried by tufts of grass. It was the carcass of a Dartmoor pony, its ribcage pecked clean by carrion, its empty eye-sockets two dark orbs. The skeleton in the foreground framed the bones of the rainforest in the distance. What had once been a living landscape had become a still-life painting. But it does not always have to be this way.

Around us, the wind sighed through the moor-grass, sounding like the whispering of restless ghosts.

Coda: If You Go Down To The Woods Today . . .

Hopefully this book has left you keen to see Britain's rainforests yourself. If you do visit a rainforest site, here are some golden rules to remember:

- When visiting temperate rainforest sites, **please treat them with the utmost respect: they're the only fragments we've got left.**
- Follow the **Countryside Code. Leave no litter, take only photographs, close farm gates if you've opened them, and keep your dog on a lead** – especially around livestock, and during nesting season for birds (March to August).
- Indeed, go further than the Countryside Code expects: **if you see that someone else has left litter, please pick it up and dispose of it.**
- **On no account damage trees, or take lichens, mosses, ferns or other plants.** Don't be like the Victorian fern-maniacs! Beware of knocking lichens, mosses and liverworts off

branches, particularly when they're dry and brittle during summer months. Many rainforests look more vibrant in the rainy winter months anyway.

- The most well-known rainforest site in Britain is Wistman's Wood, and it's starting to suffer from **visitor pressure**. If you do visit it, please don't walk through the middle of the wood – the mossy boulders are becoming overly trampled and Natural England's advice is to stick to the outside of the wood, to give the interior a break. You can see plenty of the gnarled trees and their epiphytes from the woodland edge anyway. Better still, consider making a pilgrimage to some of the many other rainforest sites discussed in this book – or hunt for as-yet-undiscovered rainforest fragments on your doorstep.

Hopefully, this book has also inspired you to become a lichen geek. If you want to identify rainforest lichens, mosses, liverworts and ferns, the following are great places to start:

- The charity **Plantlife** publishes a series of excellent free guides to identifying temperate rainforest, and the various communities of lichens, mosses and liverworts that make it special: www.plantlife.org.uk/uk/discover-wild-plants-nature/habitats/temperate-rainforest
- The websites of the **British Lichen Society** (britishlichensociety.org.uk) and **British Bryological Society** (www.britishbryologicalsociety.org.uk) contain a wealth of information on lichens, mosses and liverworts – including publications pages, where you can order the most up-to-date field guides.
- The **National Biodiversity Network Atlas** website (nbnatlas.org) is a brilliant resource for looking up botanical records of lichens and bryophyte species and their distributions.
- To see the smaller lichens and bryophytes, it's worth buying **a 10x magnification hand lens**. The National History Book Service website stocks various different brands (www.nhbs.com).

- The bryologist **Ben Averis** has produced a wonderful illustrated guide to British rainforests, 'Drizzle, Midges and Moss', downloadable at www.benandalisonaveris.co.uk/resources/.
- **Clifton Bain's** *The Rainforests of Britain and Ireland* (2015) is a great travelogue guide to many of our best rainforest sites.
- On woodlands more generally, the works of **George Peterken**, **Oliver Rackham** and **John Rodwell** are all indispensable.
- **Many botanists are on Twitter**. I recommend following the lichenologist April Windle (@aprilwindle), bryologist Ben Averis (@AverisBen), botanist Billy Fullwood (@BotanyCornwall), lichenologist Mark Powell (@obfuscans3), and the ecologist Collbradán (@collbradan).

Organisations working to protect and restore Britain's rainforests

Alliance for Scotland's Rainforest: savingscotlandsrainforest.org.uk

Celtic Rainforests Wales: celticrainforests.wales

Plantlife: www.plantlife.org.uk

The Wild Haweswater project, run by RSPB with United Utilities: wildhaweswater.co.uk

The Woodland Trust: www.woodlandtrust.org.uk

The National Trust: www.nationaltrust.org.uk

The Wildlife Trusts: www.wildlifetrusts.org

For more information on the Lost Rainforests of Britain campaign, visit lostrainforestsofbritain.org and follow me on Twitter (@guyshrubsole).

ACKNOWLEDGEMENTS

This book has many roots. My huge thanks go to the many people who nurtured it, providing it with sustenance. Firstly, to my ceaselessly generous agent, James Macdonald Lockhart, who saw its potential and helped it grow. To my brilliant editor Shoaib Rokadiya, whose judicious pruning improved this book no end. And to my publicist Katherine Patrick and the whole team at William Collins, who have brought this book into the sunlight.

My thanks to those whose pioneering work helped sow the seeds in my head for this project. To George Monbiot, for first making me aware of Britain's rainforest fragments in his book *Feral* (2013); to Clifton Bain, for writing *The Rainforests of Britain and Ireland* (2015); and to Dr Christopher Ellis, for his work to map Britain's rainforest zone. Thanks also to Joel Scott-Halkes, who coined the phrase 'bringing back Britain's rainforests'; to Clarice Holt, Annie Randall and everyone else involved in the Wild Card campaign; and to Danny Gross, terrific campaigner at Friends of the Earth, who convinced me that *yes*, Britain really does have rainforests, back when I still doubted!

It was only after moving to Devon that I really began to understand the beauty of our rainforests. My unending thanks to everyone who's made Louisa and I feel at home here: to Lewis Winks, Harriet White and their daughter Lizzie; to Kevin and Donna Cox, for being amazing hosts and great fonts of knowledge on Dartmoor's ecology; to Adrian Colston, for his wisdom about Dartmoor's commons, and for lending me loads of books; to Naomi Oakley, for pursuing a different future for farming on Dartmoor (and for introducing us to Tintin the sheep); to Tim Ferry at Moor Trees, for pioneering an alternative vision for Dartmoor's future; and to Tony Whitehead, Lisa Schneidau and the rest of the Rewilding Dartmoor group.

My thanks also to everyone who's walked and trespassed with me in Britain's rainforests: to Nick Hayes, Jon Moses, Emma Crome and her partner Guy Buckingham, Sam Lee, Robin Webster, Nigel Coles and his son Ed, Bex Trevelyan, and the rest of the Totnes Trespass Group.

Next, my thanks to the incredible botanists who've been working quietly for decades to survey and understand our temperate rainforest habitats. To Ben and Alison Averis, rockstar bryologists, endlessly generous with their time and wisdom; to Joe Hope, lichen expert-turned-farmer; to April Windle, lichenologist extraordinaire, whose knowledge of these organisms is matched only by her infectious enthusiasm; to the lichenologist Neil Sanderson, for corresponding about SSSI sites; to Rachel Jones, Alison Smith, Dave Lamacraft and others at Plantlife, for many conversations and for their excellent training workshops on identifying rainforest species; to the young Cornish botanist Billy Fullwood, for introducing me to hazel gloves fungus; to Richard Knott, ecologist at Dartmoor National Park Authority, for sharing woodland maps with me; and to Dave Bangs, trespassing botanist, for discussions about the ghyll woodlands and rainforest-like habitats of Sussex.

Our rainforests are special not only for their biodiversity, but also because of their place in our cultural imagination. My thanks to the artists, poets, historians and campaigners who have kept their memory alive: to Alan Lee, whose sublime illustrations sparked my childhood imagination, and creator of the best cover artwork an author could ever wish for; to Gwyneth

Lewis, poet, author and translator, for conversations about *The Mabinogion*, oaks, and the 'Battle of the Trees'; to Chris Rose, campaign strategist and former countryside campaigner at Friends of the Earth, for chats about the history of modern forestry; to Peter Mason, president of the Lustleigh Society, and his wife Alexa Mason, local historian, for showing us around Lustleigh Cleave and giving me access to the Society's archives; and to the staff at the Devon Heritage Centre for their help in accessing the archives of the Dartington Estate.

To all the organisations doing the hugely important work to protect and restore our rainforests: to the indefatigable Julie Stoneman at the Alliance for Scotland's Rainforest, who introduced me to countless brilliant people working on rainforest conservation projects across Scotland; to Ross McKenzie at Raleigh International, who helped organise the Scottish rainforest blessing ceremony during COP26; to Ian Dow at the Argyll and the Isles Coast and Countryside Trust (ACT); to Lee Schofield and Dave Morris at RSPB, and John Gorst at United Utilities, for showing me round Wild Haweswater and taking me to Young Wood; to Harry Barton at Devon Wildlife Trust and Alex Raeder at the National Trust, for inviting me along to Dartmoor landscape recovery meetings; to Ross Kennerley, Eleanor Lewis and Dave Rickwood at the Woodland Trust, for showing me round Fingle Woods; to Luke Barley at the National Trust for many conversations about woodland grazing, and for sending me lots of helpful journal articles; and to Phil Sturgeon and his charity, Protect Earth, for stepping in to help save High Wood.

Besides being a book, *The Lost Rainforests of Britain* is now also an ongoing campaign to protect and restore Britain's temperate rainforests. Special thanks are due to Izi Gornall, Alice Russell and Ellen Bassam at Seahorse Environmental, who saw the campaign's potential and who have worked tirelessly with me to persuade politicians to support it; to Joss Garman and Amy Mount at the European Climate Foundation, for funding the campaign; to Tim Richards, whose GIS mapping wizardry (and unending patience for my map geekery) has been invaluable in pinning down where our last rainforest fragments survive; to Fern Owen, for helping complete our crowdsourced map of rainforest fragments by uploading hundreds of people's submissions and photos; and to Ben Goldsmith, for his enthusiastic behind-the-scenes campaigning for our rainforests.

Lastly, this book is dedicated to the people who mean the most to me. To Mum and Dad, who first sparked my interest in saving the rainforests, thirty years ago. To my late grandma, Lilian Jean Maddever, whose love for nature and unfailing kindness to everyone she met makes her sorely missed. And to my partner, Louisa Casson, fearless campaigner for oceans and rainforests, and the love of my life.

LIST OF ILLUSTRATIONS

All photos taken by the author, unless otherwise stated.

Section 1:

Page 1:
Top: Wistman's Wood, Devon
Bottom left: Filmy fern (*Hymenophyllum tunbrigense*)
Bottom right: Cudbear lichen (*Ochrolechia tartarea*)

Page 2:
Map of Britain's rainforest zones (Tim Richards and Guy Shrubsole)
  (redrawn by Martin Brown)
Page 3:
Map of Britain's rainforest fragments (Tim Richards and Guy Shrubsole)
  (redrawn by Martin Brown)

Page 4:
Top: Tree lungwort (*Lobaria Pulmonaria*) (Ernst Haeckel)
Bottom: Lichen (*Lobaria scrobiculata*) (Gus Routledge)

Page 5:
Top: Sticta lichen (*Sticta sylvatica*)
Middle: 'String-of-sausages' lichen (*Usnea articulata*)
Bottom: Royal fern (*Osmunda regalis*)

Page 6:
Top: Lichen (*Degelia atlantica*) (Colin Wells)
Bottom: Dart Valley nature reserve, Devon

ENDNOTES

1. Paradise Lost

1 The Woodland Trust, 'Temperate Rainforest', https://www.woodlandtrust.org.uk/trees-woods-and-wildlife/habitats/temperate-rainforest/.

2 Dominick DellaSala (ed.), *Temperate and Boreal Rainforests of the World: Ecology and Conservation*, 2011, p.16.

3 Christopher Ellis, 'Oceanic and Temperate Rainforest Climates and Their Epiphyte Indicators in Britain', *Ecological Indicators*, 70 (2016), 125–33, https://www.researchgate.net/publication/304105743_Oceanic_and_temperate_rainforest_climates_and_their_epiphyte_indicators_in_Britain.

4 Forest Research, 'Forestry Statistics 2021', September 2021, p.14, https://cdn.forestresearch.gov.uk/2022/02/complete_fs2021_jvyjbwa.pdf

5 Quoted in Anthony Browne, 'Dimmock in Mire over Peat', *Observer*, 31 October 1999, https://www.theguardian.com/uk/1999/oct/31/anthonybrowne.theobserver1

6 Sir Charles Walker, 'We Must Do More to Protect England's Chalk Streams', *The Times*, 28 September 2020, https://www.thetimes.co.uk/article/we-must-do-more-to-protect-englands-chalk-streams-l552wbrg2.

7 Oliver Rackham, *Trees and Woodland in the British Landscape*, 1976, p. 174.

8 Clifton Bain, *The Rainforests of Britain and Ireland: A Traveller's Guide*, 2015.

9 American Geophysical Union (AGU), 'Soils in Old-Growth Treetops Can Store More Carbon than Soils under Our Feet', 14 December 2021, https://news.agu.org/press-release/soils-in-old-growth-treetops-can-store-more-carbon-than-soils-under-our-feet/. The press release summarises research presented at the AGU's fall meeting 2021 by researchers Peyton Smith and Hannah Connuck, 'Canopy Soil Capture: An Overlooked Source or Sink of Carbon in Humid Tropical Forests?', 15

December 2021, https://agu. confex.com/agu/fm21/ meetingapp.cgi/Paper/1001199.

10 Oliver Rackham, *The Ancient Woods of the Helford River*, 2019, p. 70. I am very grateful to the photographer Joe Marshall for drawing my attention to this book.

11 Wood warblers are on the red list of UK birds, meaning their declining populations are of concern to conservationists: see RSPB, 'Wood Warblers', https://www.rspb.org.uk/birds-and-wildlife/wildlife-guides/ bird-a-z/wood-warbler/. Pied flycatchers are on the amber list, meaning there is less concern about their numbers, but populations have also declined in recent years: see RSPB, 'Pied Flycatcher', https://www.rspb.org.uk/birds-and-wildlife/wildlife-guides/ bird-a-z/pied-flycatcher/, and British Trust for Ornithology, 'Pied Flycatcher *Ficedula hypoleuca*', https://app.bto.org/ birdfacts/results/bob13490.htm.

12 Gerard Manley Hopkins, 'The May Magnificat', *Poems*, 1918.

13 Hannah Ritchie and Max Roser, 'Deforestation and Forest Loss', Our World in Data, https://ourworldindata.org/ deforestation#the-world-has-lost-one-third-of-its-forests-but-an-end-of-deforestation-is-possible. 'Half of total forest loss occurred from 8,000BC to 1900; the other half occurred in the last century alone.'

14 Richard A. Houghton, 'Deforestation', in Ramesh

Sivanpillai (ed.), *Biological and Environmental Hazards, Risks, and Disasters*, 2016, pp. 313–15, https://www. sciencedirect.com/science/ article/pii/B97801239484 72000188. The area of the United Kingdom is very approximately 24 million hectares, or 240,000 square kilometres.

15 See J. R. McNeill, *Something New under the Sun: An Environmental History of the Twentieth-Century World*, 2001.

16 Dan Nosowitz, 'How the Save the Rainforest Movement Gave Rise to Modern Environmentalism', *Vox*, 16 September 2019, https://www. vox.com/the-goods/ 2019/9/16/20863152/save-the-rainforest-environmentalism-conservation.

17 The Wilderness Society, *The Franklin Blockade by the Blockaders*, 1983, p. 5.

18 The Wilderness Society, *The Franklin Blockade by the Blockaders*, 1983, p. 116.

19 NASA Earth Observatory, 'Tracking Amazon Deforestation from Above', 19 December 2019, https:// earthobservatory.nasa.gov/ images/145988/tracking-amazon-deforestation-from-above.

20 Dominick DellaSala (ed.), *Temperate and Boreal Rainforests of the World: Ecology and Conservation*, 2011, p.154.

21 Rackham, *Trees and Woodland*, p. 40. See also Rackham's account in his *Ancient*

Woodland: Its History, Vegetation and Uses in England, 2003 edition, pp. 97–9.

22 This very cursory summary draws on the much more detailed account in Rackham, *Ancient Woodland*, pp. 97–9.

23 Sir H. Godwin, *History of the British Flora: A Factual Basis for Phytogeography (Second edition)*, 1956, p. 53.

24 Frans Vera, *Grazing Ecology and Forest History*, 2000. Oliver Rackham discussed, critiqued and accepted aspects of Vera's thesis in the 2003 edition of his *Ancient Woodland*, pp. 499–506.

25 Rackham, *Ancient Woodland*, pp. 109 (map showing provinces of the Wildwood), 284 (discussion) and 481 (map showing the 'oakwood province').

26 I say this going by the extent of the temperate rainforest zone identified by Christopher Ellis, cited earlier.

27 I. G. Simmons, 'Pollen Diagrams from Dartmoor', *New Phytologist*, 63:2 (1964), 165–80, https://nph. onlinelibrary.wiley.com/doi/ pdf/10.1111/j.1469-8137.1964. tb07369.x. Simmons's pollen diagrams at the end of the article include results for *Polypodium* (common polypody fern). There is an ongoing academic debate about how far the peat bogs that dominate Britain's uplands today were created as a result of human deforestation or because of changes in the climate during

this period. It seems likely the two interacted: to Simmons, 'it seems inescapable that at least some of the early recession of forest which allowed blanket bog growth was due to man's activity'. Once the waterlogged, acidic peat bogs had formed, natural regeneration of trees on the uplands became harder.

28 D. J. Maguire and C. J. Caseldine, 'The Former Distribution of Forest and Moorland on Northern Dartmoor', *Area*, Vol. 17, No. 3 (Sep., 1985), p. 193.

29 Ibid., p. 201, table 4: 'The estimated height of the tree line at various locations in the British Isles during the Flandrian', where the figure given for the Lake District is 760 metres. The Atlantic period was the most favourable climatic phase of the Flandrian for forest cover.

30 Andrew Weatherall et al., 'Young Wood: A Woodland beyond the Edge', in Ian D. Rotherham et al. (eds), *Trees beyond the Wood: An Exploration of Concepts of Woods, Forests and Trees*, 2012, p. 313. Available online at https://www. researchgate.net/ publication/272784223_Young_ Wood_a_woodland_beyond_ the_edge.

31 Rackham, *Trees and Woodland*, 1976, pp. 45–7.

32 E. L. Kellogg, *Coastal Temperate Rain Forests: Ecological Characteristics, Status, and Distribution Worldwide*, 1992.

Available online at http://
archive.ecotrust.org/
publications/ctrf.html.

33 Aldo Leopold, *A Sand County
Almanac,* 1949.

34 George Monbiot, *Feral:
Searching for Enchantment on
the Frontiers of Rewilding,* 2013,
pp. 67–8.

35 Bain, *Rainforests of Britain and
Ireland.*

36 DellaSala (ed.), *Temperate and
Boreal Rainforests,* p. 154.

37 The Google Map I made can
be seen here: https://www.
google.com/maps/d/
edit?mid=1VX8n1mzcl
N2OGSPTPUtM2r2
txKOceDEz&usp=sharing

38 Guy Shrubsole, 'Life finds a
way: in search of England's lost,
forgotten rainforests', *Guardian,*
29 April 2021, https://www.
theguardian.com/
environment/2021/apr/29/life-
finds-a-way-in-search-of-
englands-lost-forgotten-
rainforests.

2. Ghosts in the Landscape

1 Conan Doyle wrote these
words in a letter dated 2 April
1901 to his mother from the
Old Duchy Hotel in
Princetown. See Simon Calder,
'Where Sherlock Holmes
Feared to Tread', *Independent,* 2
March 1996, https://www.
independent.co.uk/travel/where-
sherlock-holmes-feared-to-
tread-1339829.html.

2 Quoted in W. W. Robson's
introduction to the Oxford
World's Classics edition of
Arthur Conan Doyle, *The
Hound of the Baskervilles,* 1998,
p. xii.

3 For more on the Dartmoor
places that inspired *The Hound
of the Baskervilles,* see Philip
Weller, *The Hound of the
Baskervilles: Hunting the
Dartmoor Legend,* 2001, and
BBC Devon, http://www.bbc.
co.uk/devon/outdoors/moors/
hound_baskervilles.shtml.

4 Wistman's Wood is first
mentioned by Tristram Risdon,
whose *Survey of the County of
Devon* was written around
1620, although it was not
published until later. A
summary of tree-ring and
growth-rate studies of the trees
in Wistman's Wood can be
found in M. C. F. Proctor, G.
M. Spooner and M. Spooner,
'Changes in Wistman's Wood,
Dartmoor: Photographic and
Other Evidence', *Report and
Transactions of the Devonshire
Association for the Advancement
of Science, Literature and the
Arts,* 112 (1980), table on p.
68.

5 Jennifer Westwood, *Albion: A
Guide to Legendary Britain,*
1985, p. 32; Eric Hemery,
*High Dartmoor: Land and
People,* 1983, pp. 454–5.

6 John Fowles, *The Tree,* 1978.

7 I. G. Simmons, 'The Dartmoor
Oak Copses: Observations and
Speculations', *Field Studies,* 2
(1965), pp. 225–35, cited by
E. P. Mountford, P. A. Page
and G. F. Peterken, 'Twenty-
Five Years of Change in a
Population of Oak Saplings in
Wistman's Wood, Devon',

English Nature Research Reports No. 348, 2000. Available online at http://publications.naturalengland.org.uk/publication/215866.

8 See, for example, the old saying recounted in 'Under Bracken . . .?', Plant-Lore, 10 February 2015, https://www.plant-lore.com/news/under-bracken/.

9 Ian D. Rotherham, 'Preface', in Ian D. Rotherham and Christine Handley (eds), *Shadow Woods and Ghosts: A Survey Guide*, 2013, p. 1.

10 Adrian Colston, 'The Oak Brook – A Place of Mutual Satisfaction?', A Dartmoor Blog, 22 March 2016, https://adriancolston.wordpress.com/2016/03/22/the-oak-brook-a-place-of-mutual-satisfaction/.

11 Horace Waddington, 'Straight across Dartmoor', *Exeter and Plymouth Gazette*, 11 January 1867.

12 See John Page, *An Exploration of Dartmoor and its Antiquities*, 1895, p. 103, and Tim Sandles, 'Watern Oak', Legendary Dartmoor, 19 March 2016, https://www.legendarydartmoor.co.uk/watern_oke.htm. Horace Waddington, in the *Exeter and Plymouth Gazette* on 11 January 1867, also saw the Watern Oak whilst it was still standing. Having climbed to the top of Amicombe Hill, he observes 'southwards, on our left, the infant Tavy winding through Tavy Wane, and bound for ancient Tavistock; beyond it, Watern Oak, a solitary tree, backed by the abrupt crags and lofty peak of Fur Tor'. This description, and the 1809 Ordnance Survey map, appear to put the Watern Oak to the southeast of where the modern OS map records 'Watern Oke'.

13 William Crossing, *Guide to Dartmoor*, 1909, pp. 175–6.

14 The name may also mean 'Far Tor' (given its remoteness) or 'Great Tor' (given its height). See Tim Sandles, 'Fur Tor', Legendary Dartmoor, 30 March 2016, https://www.legendarydartmoor.co.uk/queen_moor.htm.

15 Crossing, *Guide to Dartmoor*, pp. 39–40. See also L. A. Harvey and D. St Leger-Gordon, *Dartmoor* (Collins New Naturalist Series), 1953, p. 181: 'The Forest [of Dartmoor] was reserved as such mainly for the maintenance of deer, with the essential cover – hence the "verte and venison" proviso.'

16 Horace Waddington, in the *Exeter and Plymouth Gazette* on 11 January 1867, states that Dunnagoat means 'Dun-a-coet (Saxon for "underwood")'. The geologist and historian R. N. Worth, in 'Notes on the Historical Connections of Devonshire Place-Names', *Report and Transactions of the Devonshire Association for the Advancement of Science, Literature and the Arts*, 10 (1878), p. 283, has this assessment: 'Dunnagoat, the final syllable of which is clearly the Kornu [Cornish Celtic]

coed, "a wood"; whilst in *dun* we have "hill" = *dun-y-coed*, "the wooded hill".' Crossing, in his *Guide to Dartmoor*, p. 180, relates: 'Mr Richard John King says that Dunnagoat, or Danagoat, as it is sometimes spelt, is "from the Cornish *dan*, *under*, and *coet, a wood*". He takes the name to belong to a hollow. But we incline to think it more probable that in the first syllable of the name we see the Celtic *dun*, a hill, and if the second really is *coet*, or *coed*, that this may have been derived from the former presence of trees in the valley of the Rattle Brook. Even now in parts of it a solitary rowan, or oak, is to be met with.'

17 Duchy of Cornwall, 'Dartmoor & Princetown', https:// duchyofcornwall.org/dartmoor-and-princetown.html.

18 Alice Oswald, *Dart*, 2002, p. 11. See also Ian Mercer, *Dartmoor* (Collins New Naturalist Series), 2009, p. 141: 'The word "dart" . . . is a derivative of the British "daerwint" (and "derw" is oak) and mean oak tree river in the Celtic languages that almost always gave the British landscape its original name for physical features.'

19 Pliny the Elder, *Natural History* (70s AD) book 16, chapter 95, trans. John Bostock, 1855, https://www.perseus.tufts.edu/ hopper/text. jsp?doc=Perseus%3Atext% 3A1999.02. 0137%3 Abook%3 D16%3Achapter%3D95.

20 Tim Sandles, 'Druids', Legendary Dartmoor, 16 March 2016, https://www. legendarydartmoor.co.uk/druid_dawn.htm.

21 Crossing, *Guide to Dartmoor*, p. 18.

22 For an online Google Map of the various place names indicating possible ghost woods related in this chapter, see the one I've made at https://www. google.com/maps/d/u/0/ edit?mid=16qcfr9F2ykH Wezre84PSGqo5ARDX rnW4&usp=sharing.

23 C. E. Stevens, 'The Sacred Wood', in J. V. S. Megaw (ed.), *To Illustrate the Monuments: Essays on Archaeology Presented to Stuart Piggott*, 1976. See also Kelly A. Kirkpatrick, 'Nemeton in the Medieval World', Nemi to Nottingham: Following in the Footsteps of Fundilia, 3 September 2013, https:// nemitonottingham.wordpress. com/2013/09/03/nemeton-in-the-medieval-world/.

24 Ronald Hutton, *Pagan Britain*, 2013, pp. 216–17.

25 Tithe maps and apportionments for the parish of Lydford (Forest of Dartmoor) are available to view on the Devon County Council website, https://www.devon.gov.uk/ historicenvironment/tithe-map/ lydford-forest-of-dartmoor/. Field parcels 1065,1070 and 1071 are entitled 'The Bearas'. I'm very grateful to Peter and Alexa Mason for suggesting investigating tithe maps for old place names indicating former

woods. Crossing, *Guide to Dartmoor*, p. 466, also records that at Week Ford, a crossing over the Dart on the edge of these fields, there is 'an old blowing-house [a building in which tin was smelted], in which an oak is now growing. The building is called Beara House in the locality'.

26 Crossing, *Guide to Dartmoor*, p. 196.

27 Documents cited in J. P. Barkham, 'Pedunculate Oak Woodland in a Severe Environment: Black Tor Copse, Dartmoor', *Journal of Ecology*, 66:3 (1978), pp. 707–40.

28 Lady Raglan, 'The "Green Man" in Church Architecture', *Folklore*, 50:1 (1939), pp. 45–57.

29 Hutton, *Pagan Britain*, discusses the Green Man on pp. 347–51, concluding that after 'sustained scholarly investigation of the pagan interpretation . . . the interpretation rapidly collapsed'.

30 Adrian Cooper, director of the charity Common Ground (which maintains one of the largest collections of Green Man images, compiled by the late botanist Kathleen Basford), told art critic Alastair Sooke that he thought 'the Green Man emerged from pre-Christian times, out of the wild woods, really': 'The Surprising Roots of the Mysterious Green Man', BBC Culture, 4 January 2019, https://www.bbc.com/culture/article/20190104-the-surprising-roots-of-the-mysterious-green-man. The musician Mike Harding, who has also amassed a large collection of Green Man sightings, notes: 'The only pattern I have found so far is that he seems to appear in his greatest concentrations in Europe wherever there are stretches of old relict woodlands. Thus the biggest collections I have discovered so far seem to be in Devon and Somerset': 'Stories in the Leaves', The Mystery of the Green Man, https://www.mikeharding.co.uk/greenman/stories-in-the-leaves/.

31 Author's visit to St Pancras Church, 2021, and A4 guide published by the church giving details of the carved bosses on the chancel ceiling.

32 Dominick DellaSala (ed.), *Temperate and Boreal Rainforests of the World: Ecology and Conservation*, 2011, p. 245.

33 Noel Carrington, *Dartmoor: A Descriptive Poem*, 1826.

34 Crossing, *Guide to Dartmoor*, p. 20.

35 Miller Christy and R. H. Worth, 'The Ancient Dwarfed Oak Woods of Dartmoor', *Report and Transactions of the Devonshire Association for the Advancement of Science, Literature and the Arts*, 54 (1922), https://devonassoc.org.uk/wd/wd074.pdf.

36 HMSO, *Dartmoor: National Park Guides Number One*, 1957, pp. 14–15.

37 Natural England, 'Devon's National Nature Reserves', 13

May 2001, https://www.gov.uk/government/publications/devons-national-nature-reserves/devons-national-nature-reserves#wistmans-wood.

38 L. A. Harvey and D. St Leger-Gordon, *Dartmoor* (Collins New Naturalist Series), 3rd edition, 1977, pp. 235–6.

39 Mercer, *Dartmoor*, p. 164.

40 Dartmoor Trust archives, https://dartmoortrust.org/archive/record/16376.

41 Proctor, Spooner and Spooner, 'Changes in Wistman's Wood, Dartmoor: Photographic and Other Evidence'. Rep. Trans. Devon. Ass. Advmt. Sci. 112, 43–79, December 1980. I am grateful to the Devonshire Association for sending me a scanned copy of this paper. See also https://devonassoc.org.uk/person/spooner-g-m/.

42 Ibid, p. 44.

43 E. P. Mountford, C. E. Backmeroff and G. F. Peterken, 'Long-Term Patterns of Growth, Mortality, Regeneration and Natural Disturbance in Wistman's Wood, a High Altitude Oakwood on Dartmoor', *Transactions of the Devonshire Association for the Advancement of Science*, 133 (2001), pp. 231–62.

44 Robert Burnard's map of 'Plundered Dartmoor' suggests Wistman's Wood (Longaford Newtake) wasn't enclosed until after 1820, but before 1895. J. F. Archibald's short paper 'Wistman's Wood: A Forest Nature Reserve', *Journal of the Devon Trust for Nature Conservation*, 11 (December 1966), pp. 423–8, states that Wistman's Wood was enclosed as part of Longaford Newtake in 1818. Thanks to Luke Barley at the National Trust for sending me a copy of this paper.

45 Sabine Baring-Gould, *A Book of the West, Being an Introduction to Devon and Cornwall*, 1899, vol. I, pp. 182–3, https://en.wikisource.org/wiki/A_Book_of_the_West/Volume_1/11.

46 See also Seán Jennett, *Deserts of England*, 1964, p. 176, regarding the West Dart: 'On the right bank is Beardown Tor and Beardown Hill, a name derived, not from the presence of bears, but from a word meaning "wood" that is still to be found on Dartmoor, notably in the name Black Tor Beare.'

3. Trespassing Botanists & Fern Maniacs

1 See the Right to Roam campaign website: https://www.righttoroam.org.uk/.

2 William Crossing, *Guide to Dartmoor* (1909), pp. 340–1.

3 Fungi, which form an entirely separate kingdom of life to plants, are also considered cryptogams, as they reproduce using spores.

4 British Pteridological Society, 'Introduction to Ferns', https://ebps.org.uk/ferns/identification/key-to-common-british-native-ferns/introduction-to-ferns/.

5 JoAnna Klein, 'In the Race to

Live on Land, Lichens Didn't Beat Plants', *New York Times*, 19 November 2019, https://www.nytimes.com/2019/11/19/science/lichens-plants-evolution.html.

6 Charles C. Plitt, 'A Short History of Lichenology', *The Bryologist*, 22:6 (1919), p. 77, https://www.jstor.org/stable/3238526?seq=1#metadata_info_tab_contents.

7 Robin Wall Kimmerer, *Gathering Moss: A Natural and Cultural History of Mosses*, 2003, p. ix.

8 Plantlife, 'Lichens and Bryophytes of Atlantic Woodland in Scotland: An Introduction to Their Ecology and Management', 2010, p. 6, https://www.plantlife.org.uk/application/files/9914/8233/4028/PLINKS_AtlanticWoodland_LRes.pdf.

9 Derek Ratcliffe, 'An Ecological Account of Atlantic Bryophytes in the British Isles', *New Phytologist*, 67 (1968), pp. 365–439, https://www.jstor.org/stable/2430430.

10 Michael Stech et al., 'Bryophytes and Lichens in 16th-Century Herbaria', *Journal of Bryology*, 40:2 (2018), pp. 1–8, https://www.researchgate.net/publication/324596633_Bryophytes_and_lichens_in_16th-century_herbaria.

11 Stuart D. Crawford, 'Lichens Used in Traditional Medicine', in Branislav Ranković (ed.), *Lichen Secondary Metabolites: Bioactive Properties and Pharmaceutical Potential*, 2015,

pp. 27–80, https://arctichealth.org/media/pubs/297033/9783319133737-c2.pdf.

12 National Biodiversity Network Atlas: https://nbnatlas.org/.

13 Woodland Trust, 'Wild Service Tree', https://www.woodlandtrust.org.uk/trees-woods-and-wildlife/british-trees/a-z-of-british-trees/wild-service-tree/.

14 Cecilia Ronnås et al., 'Discovery of Long-Distance Gamete Dispersal in a Lichen-Forming Ascomycete', *The New Phytologist*, 216:1 (2017): 216–26, https://www.ncbi.nlm.nih.gov/pmc/articles/PMC5655791/.

15 Francis Rose, 'Ancient British Woodlands and Their Epiphytes', *British Wildlife*, 5:2 (1993), pp. 83–93, https://www.britishwildlife.com/article/volume-5-number-2-page-83-93.

16 National Trust, 'Charity Attempts Largest Transplant of Ancient and Rare Lichen in Efforts to Protect Its Future', 20 November 2020, https://www.nationaltrust.org.uk/press-release/charity-attempts-largest-transplant-of-ancient-and-rare-lichen-in-efforts-to-protect-its-future-.

17 Trees for Life, 'Caledonian Forest – Species Profile: Tree Lungwort', 2008, https://treesforlife.org.uk/archive/members/20081127mr/Tree_lungwort.pdf.

18 As Francis Rose says: 'It has been suggested that the

[Lobarion] community is a very oceanic one, but old records and herbarium specimens make it clear that, before air pollution became extensive, it occurred very widely through mid-England, eastwards into Cambridgeshire, Suffolk, Essex and Lincolnshire'. See Rose, 'Ancient British Woodlands and their Epiphytes', p. 88.

19 Mark Pyron, 'Characterizing Communities', *Nature Education Knowledge* 3(10):39 (2010), https://www.nature.com/scitable/knowledge/library/characterizing-communities-13241173/.

20 Crawford, 'Lichens Used in Traditional Medicine', p. 62. *Peltigera canina* is widespread across Britain, but two rainforest species of *Peltigera* are *Peltigera collina* (which has floury edges) and *Peltigera horizontalis* (with chestnut-brown fruit that stick out at right angles from the trunk).

21 Dictionaries of the Scots Language, 'Crottle *n.1*', 2004, https://www.dsl.ac.uk/entry/snd/crottle_n1.

22 For help identifying these lichens and many more, see Plantlife's guides to the Lobarion and Parmelion lichens, available to download at https://www.plantlife.org.uk/uk/our-work/publications. For a more detailed overview of the various different lichen communities, see Peter W. James, David L. Hawksworth and Francis Rose, 'Lichen Communities in the British Isles: A Preliminary Conspectus', in Mark R. D. Seaward (ed.), *Lichen Ecology*, 1977, pp. 295–413, https://www.britishlichensociety.org.uk/sites/www.britishlichensociety.org.uk/files/pdf/Lichen%20Communities%20Complete.pdf.

23 Quoted in Nicola Chester, 'All about Lichens', RSPB Natures Home Magazine Uncovered, 14 February 2019, https://community.rspb.org.uk/ourwork/b/natureshomemagazine/posts/all-about-lichens.

24 Vasudeo P. Zambare and Lew P. Christopher, 'Biopharmaceutical Potential of Lichens', *Pharmaceutical Biology*, 50:6 (2012), 50:6, p. 778, https://www.tandfonline.com/doi/full/10.3109/13880209.2011.633089.

25 See the Ashmolean Museum's webpage for Ruskin's drawing: http://ruskin.ashmolean.org/collection/9006/9037/9356/all/per_page/25/offset/0/sort_by/seqn./object/14350.

26 The drawing is reproduced in Lilias Wigan, 'In Focus: The Exquisite yet Tiny Botanical Study by John Ruskin that Took Him Four Years to Complete', *Country Life*, 5 April 2019, https://www.countrylife.co.uk/luxury/art-and-antiques/focus-exquisite-yet-tiny-botanical-study-john-ruskin-took-four-years-complete-194499.

27 A brilliant essay on this subject

is Kate Flint, 'Ruskin and Lichens', in Kelly Freeman and Thomas Hughes (eds), *Ruskin's Ecologies: Figures of Relation from Modern Painters to The Storm-Cloud*, 2021, https://courtauld.ac.uk/research/research-resources/publications/courtauld-books-online/ruskins-ecologies-figures-of-relation-from-modern-painters-to-the-storm-cloud/1-ruskin-and-lichen-kate-flint/.

28 A. M. Averis, A. B. G. Averis et al., *An Illustrated Guide to British Upland Vegetation*, 2004, https://hub.jncc.gov.uk/assets/a17ab353-f5be-49ea-98f1-8633229779a1. With characteristic modesty, Ben and Alison were embarrassed by my description of their work as 'the' book on upland vegetation, and asked that I mention a preceding seminal work that inspired them: Donald N. McVean and Derek A. Ratcliffe, *Plant Communities of the Scottish Highlands: A Study of Scottish Mountain, Moorland and Forest Vegetation*, 1962.

29 For a more detailed comparison, see the British Bryological Society's field guide accounts for *Plagiochila spinulosa* (https://www.britishbryologicalsociety.org.uk/wp-content/uploads/2020/12/Plagiochila-spinulosa.pdf) and *P. punctata* (https://www.britishbryologicalsociety.org.uk/wp-content/uploads/2020/12/Plagiochila-punctata.pdf).

30 Kimmerer, *Gathering Moss*, p. 10.

31 Sarah Whittingham, *The Victorian Fern Craze*, 2009, p. 9.

32 Plantlife, 'Ferns of Atlantic Woodland', 2020, https://www.plantlife.org.uk/application/files/3915/9438/8641/Ferns_of_Atlantic_Woodlands_WEB.pdf.

33 Charlotte Kingsley Chanter, *Ferny Combes: A Ramble after Ferns in the Glens and Valleys of Devonshire*, 1856, pp. 68 and 64, https://www.google.co.uk/books/edition/Ferny_Combes/lPXapkvfvm0C?hl=en&gbpv=1.

34 Francis George Heath, *The Fern Paradise: A Plea for the Culture of Ferns*, 1878 edition, p. 54, https://books.google.co.uk/books?id=R0kDAAAAQAAJ&printsec=frontcover&source=gbs_ge_summary_r&cad=0#v=onepage&q&f=false.

35 Whittingham, *Victorian Fern Craze*, pp. 11–12.

36 For the fern designs in Oxford's Natural History Museum, see John Holmes, 'Rebels of Arts and Science: The Empirical Drive of the Pre-Raphaelites', *Nature*, 562, 25 October 2018, pp. 490–1, https://www.nature.com/articles/d41586-018-07110-9. On custard creams and fernmania, see Dimitra Nikolaidou, 'How the Victorian Fern-Hunting Craze Led to Adventure, Romance, and Crime', Atlas Obscura, 12 December 2016, https://www.atlasobscura.com/articles/how-the-victorian-fern-hunting-craze-led-to-adventure-romance-and-crime.

37 Whittingham, *Victorian Fern Craze*, p. 14.

38 Frederick Hamilton Davey, *Flora of Cornwall, Being an Account of the Flowering Plants and Ferns Found in the County of Cornwall, Including the Scilly Isles*, 1909.

39 Francis Brent, 'The Botany of Dartmoor and Its Borders', in Samuel Rowe, *A Perambulation of the Antient and Royal Forest of Dartmoor and the Venville Precincts*, 1896 edition, p. 352. For more on the destructiveness of fernmania, see P. D. A. Boyd, 'The Victorian Fern Cult in South-West Britain', in J. M. Ide, A. C. Jermy and A. M. Paul (eds), *Fern Horticulture: Past, Present and Future Perspectives*, 1992, 33–56, http://www.peterboyd.com/ferncultsw.htm.

40 Nona Bellairs, *Hardy Ferns: How I Collected and Cultivated Them*, 1865, p. 77.

41 Both examples cited in Whittingham, *Victorian Fern Craze*, p. 19.

42 Ward Lock & Co., *Guide to North Wales (Southern Section)*, 1925, p. 99. The warning can be found under the entry for Cwm Bychan Lake in the Rhinog mountains, a spot where temperate rainforest still thrives today.

43 Devon's 1906 Fern Law is mentioned in Sarah Whittingham, *Victorian Fern Craze*, p. 53, but the full wording is quoted in Boyd, 'Victorian Fern Cult in South-West Britain'. To calculate the value of £5 in 1906 in today's money, I used the UK Inflation Calculator at https://www.officialdata.org/UK-inflation.

44 Robert Michael Pyle, 'The Extinction of Experience', in Terrell F. Dixon (ed.), *City Wilds: Essays and Stories about Urban Nature*, 2002, p. 263.

4. Seeing the Wood for the Trees

1 Andrew Weatherall et al., 'Young Wood: A Woodland beyond the Edge', in Ian D. Rotherham et al. (eds), *Trees beyond the Wood: An Exploration of Concepts of Woods, Forests and Trees*, 2012, https://insight.cumbria.ac.uk/id/eprint/1522/1/Weatherall_YoungWood.pdf. The UK government's National Forest Inventory, the official map of woodland cover in Britain, states that the minimum size for a wood to be included is 0.5 hectares (1.25 acres): see https://www.gov.uk/government/statistics/national-forest-inventory-tree-cover-outside-woodland-in-gb.

2 The two global distribution maps of temperate rainforest I have been able to find are in E. L. Kellogg, *Coastal Temperate Rain Forests: Ecological Characteristics, Status, and Distribution Worldwide*, 1992 (available online at http://archive.ecotrust.org/publications/ctrf.html), and Dominick DellaSala (ed.), *Temperate and Boreal Rainforests of the World: Ecology and Conservation*, 2011.

3 DellaSala (ed.), *Temperate and Boreal Rainforests*, p. 154.

4 Ben Averis, personal communication with the author, 2021.

5 George Monbiot, *Feral: Searching for Enchantment on the Frontiers of Rewilding*, 2013, p. 67.

6 Clifton Bain, *The Rainforests of Britain and Ireland: A Traveller's Guide*, 2015, p. 59: 'The 38 sites presented in this book have been selected to include those woods with reasonable access and to represent the geographical spread and characteristics of Atlantic rainforests in Britain and Ireland.'

7 Christopher Ellis, 'Oceanic and Temperate Rainforest Climates and Their Epiphyte Indicators in Britain', *Ecological Indicators*, 70 (2016), 125–33, https://www.researchgate.net/publication/304105743_Oceanic_and_temperate_rainforest_climates_and_their_epiphyte_indicators_in_Britain.

8 See also Oliver Rackham's posthumously published book *The Ancient Woods of the Helford River*, 2019.

9 Paul B. Alaback, 'Comparative Ecology of Temperate Rainforests of the Americas along Analogous Climatic Gradients', *Revista Chilena de Historia Natural*, 64 (1991), 399–412, http://rchn.biologiachile.cl/pdfs/1991/3/Alaback_1991.pdf.

10 UK Met Office, average annual rainfall records for the weather monitoring station at Hampstead, north London, 1991–2020: https://www.metoffice.gov.uk/research/climate/maps-and-data/uk-climate-averages/gcpv7fnqu; and for Woodford, Greater Manchester, 1991–2000: https://www.metoffice.gov.uk/research/climate/maps-and-data/uk-climate-averages/gcqrqyr80.

11 Ellis, 'Oceanic and Temperate Rainforest Climates', p.126.

12 Chris Ellis, personal communication with the author, January 2021.

13 Ben Averis, personal communication with the author, January 2022.

14 A. G. Tansley, *Britain's Green Mantle: Past, Present and Future*, 1949, p. 46.

15 A. G. Tansley, *The British Islands and Their Vegetation*, 1939, vol. I, p. 55.

16 J. J. Amann, 'L'hygrothermie du climat, facteur déterminant la répartition des espèces atlantiques', *Revue bryologique et lichénologique*, 2 (1929), 126–33. Amann's index is also discussed in P. Greig-Smith, 'Evidence from Hepatics on the History of the British Flora', *Journal of Ecology*, 38:2 (1950), pp. 320–44, https://www.jstor.org/stable/2256449. The full equation for determining Amann's index of hygrothermy (H) is as follows: $H = [(P \times T)/t_h - t_c]$, where P is the mean annual precipitation, T is the mean annual temperature, and t_h and t_c are the mean temperatures of the warmest

and coldest months, respectively.

17 M. C. F. Proctor, 'Mosses and Liverworts of the Malham District', *Field Studies*, 1:2 (1960), p. 61, https://fsj.field-studies-council.org/media/344693/vol1.2_13.pdf.

18 As Greig-Smith, 'Evidence from Hepatics', states, p. 336: 'Unfortunately, data are not available for considerable stretches of the west coasts of Britain and Ireland, where the main concentrations of the Atlantic species occur.'

19 Ben Averis, personal communication with the author.

20 See the work of Ralph W. V. Elliot to situate the Green Knight in a geographical setting: 'Sir Gawain in Staffordshire: A Detective Essay in Literary Geography', *Times*, 21 May 1958. Thanks to the artist George Outhwaite for bringing Lud's Church to my attention and suggesting it as a potential rainforest microclimate.

21 For an overview of the seemingly limited amount of research done into the role of occult precipitation as a water source for British sites, see L. A. Sampurno Bruijnzeel, Werner Eugster and Reto Burkard, 'Fog as a Hydrologic Input', in *Encyclopedia of Hydrological Sciences*, 2006, https://www.researchgate.net/publication/230223404_Fog_as_a_Hydrologic_Input.

22 To get a better sense of the legend that is Dave Bangs, see Jonathan Moses, 'The Need to Trespass: Let People in to Protect Nature, Says Guerrilla Botanist', *Guardian*, 26 November 2021, https://www.theguardian.com/environment/2021/nov/26/david-bangs-sussex-guerrilla-botanist-trespass-protecting-nature-aoe. Dave Bangs's book *The Land of the Brighton Line: A Field Guide to the Middle Sussex and South East Surrey Weald*, 2018, is a magisterial book, weaving together botany, history and land ownership, with a side-order of anti-capitalist polemic.

23 Francis Rose et al., *Atlas of Sussex Mosses, Liverworts and Lichens*, 1991, pp. 20–4. Thanks to Dave Bangs for sending me a copy of the relevant section in this publication. Another useful study of ghyll woodland is Niall G. Burnside et al., 'Ghyll Woodlands of the Weald: Characterisation and Conservation', *Biodiversity and Conservation*, 15 (2006), 1319–38, https://www.researchgate.net/publication/226723829_Ghyll_Woodlands_of_the_Weald_Characterisation_and_Conservation.

24 For example, see the Sussex Wildlife Trust's webpage 'Wet Woodland', with its entry for ghyll woodland: 'This rare habitat type is a unique landscape feature of this part of Sussex and of the UK.' Nevertheless, it also states: 'The

flora found in these sites is very characteristic of former Atlantic conditions – including lush growths of ferns (such as hay scented buckler fern), mosses and liverworts' – https://sussexwildlifetrust.org.uk/discover/around-sussex/wetland-habitats/wet-woodland.

25 Oliver Rackham, *Ancient Woodland: Its History, Vegetation and Uses in England*, 2003 edition, pp. 124–5. Concerning the Weald, Rackham concludes that 'roughly 70% of the area, some 600,000 acres, was woodland in Domesday times'.

26 Stef Haesen et al., 'ForestTemp – Sub-Canopy Microclimate Temperatures of European Forests', *Global Change Biology*, 27:23 (2021), pp. 6307–19, https://onlinelibrary.wiley.com/doi/abs/10.1111/gcb.15892. Thanks to Professor Ilya Maclean at Exeter University for flagging this paper to me.

27 L. A. Harvey and D. St Leger-Gordon, *Dartmoor* (Collins New Naturalist Series), 1953, p. 129.

28 George Peterken, *Woodland Conservation and Management*, 1993 edition, p. 107.

29 Ibid., pp. 107–16. Peterken's typology of birch–oak woodland ('Group 6') can be found on pp. 141–8.

30 A good overview of the National Vegetation Classification can be found on the website of the Joint Nature Coordinating Committee (JNCC): https://jncc.gov.uk/our-work/nvc/. The key text for NVC woodlands is J. S. Rodwell (ed.), *British Plant Communities, Volume 1: Woodlands and Scrub*, 1991.

31 See Peterken, *Woodland Conservation and Management*, p. 141, and Rodwell, (ed.), *British Plant Communities*, pp. 284–5.

32 Peterken, *Woodland Conservation and Management*, p. 320, notes that 'the NVC . . . excludes epiphytes'. Epiphytic lichens and mosses are certainly *mentioned* by Rodwell's *British Plant Communities*, but do not tend to play a role in its categorisation of woodland types. The bryologist Ben Averis, who uses the NVC system for his surveys, explained to me that 'the reason why lichens don't feature as much as bryophytes in the NVC is because the NVC is based on quadrat samples . . . of ground vegetation. Some epiphytes might have been included in some of those quadrats but generally the sampling was of ground habitats, including rocks in some cases. On the woodland floor bryophytes are generally much more abundant than lichens.'

33 NVC maps covering much of Scotland can be downloaded from the NatureScot website: https://cagmap.snh.gov.uk/natural-spaces/dataset.jsp?code=NVC.

34 John Rodwell, personal communication with the

author. Another problem is that most NVC survey maps will have been paper versions drawn by ecologists in the 1990s, and only some will have been digitised subsequently. I also asked the woodland ecologist Keith Kirby whether NVC polygon maps existed for all of Britain's woods. He told me: 'The short answer is that I don't think there are any country-wide polygon maps. They would really require field surveys. Only a proportion of woods have been mapped to that level, particularly in England (more done in Wales and Scotland) and I am not aware of any central repository for collating such surveys as are done.' He added: 'I am not surprised you are getting some odd results with overlaying the NVC points with NFI [National Forest Inventory] polygons – some of the points will be approximations when a type was known to be present in a wood or even a 10km square, but no more precise record; also a lot of records will be pre-1990 so the woods might have been felled or replanted since.'

35 Oliver Rackham, South Helford River 1 notebook, 1980–90, p. 2, Corpus Christi College Cambridge Archive, CCCC14/6/2/2/ HelfordRiverSouth1, https:// cudl.lib.cam.ac.uk/view/ MS-CCCC-00014-00006- 00002-00002-HELFORDR IVERSOUTH-00001/1.

36 Rackham, *Ancient Woodland*, p. 511.

37 To arrive at this figure, Tim Richards downloaded all NBN Atlas records for a set of oceanic indicator species of lichens, mosses and liverworts, based on a list provided to us by Ben Averis. Each record has a grid reference location, although some of these locations are not precise; records with only fuzzy geolocations (e.g. hectad level) were removed. The remaining records were then intersected with our map of ancient woodland in Britain's oceanic zone. The resulting area of ancient woods in the oceanic zone which had at least one oceanic indicator species of lichen or bryophyte present was 34,730 hectares (86,000 acres), or 25.8 per cent of the total area of ancient woodland in the oceanic zone (333,000 acres).

38 The website homepage of the Alliance for Scotland's Rainforests (https:// savingscotlandsrainforest.org. uk/) states that as little as 30,000 hectares (i.e. 74,000 acres) of temperate rainforest remain in Scotland. My thanks to Plantlife for sending me the methodology and maps that underpin these calculations, which deployed a similar approach to the one that Tim Richards and I did to map rainforests across Britain as a whole.

5. Myth, Magic and *The Mabinogion*

1 Celtic Rainforests Wales, 'Coed Felenrhyd / Llennyrch', https://celticrainforests.wales/coed-felenrhyd-llennyrch. See also Woodland Trust, 'Coed Felenrhyd & Llennyrch', https://www.woodlandtrust.org.uk/visiting-woods/woods/coed-felenrhyd-llennyrch/.

2 Gwynedd Archaeological Trust states the height of the Trawsfynydd dam to be ninety-six feet: 'Historic Landscape Characterisation: Trawsfynydd – Area 5: Trawsfynydd Power Station and Lake', http://www.heneb.co.uk/hlc/trawsfynydd/traws5.html.

3 *The Mabinogion*, trans. Gwyn Jones and Thomas Jones, 1948, p. ix.

4 *The Mabinogion*, trans. Sioned Davies, 2007, p. x.

5 *The Book of Taliesin: Poems of Warfare and Praise in an Enchanted Britain*, trans. Gwyneth Lewis and Rowan Williams, 2020, p. 202.

6 Tacitus, *Annals* (early second century AD), book 14, Chapter 30, trans./ A. J. Church and W. J. Brodribb, 1969, https://www.perseus.tufts.edu/hopper/text?doc=Perseus%3Atext%3A1999.02.0078%3Abook%3D14%3Achapter%3D3. Julius Caesar twice invaded Britain unsuccessfully, in 55 and 54 BC. Britain was successfully invaded again by the Romans in AD 43, but their forces did not reach Anglesey until around AD 60.

7 *The Mabinogion*, trans. Davies, p. 51.

8 'The Stanzas of the Graves' (from the thirteenth-century *Black Book of Carmarthen*) records that 'In Aber Gwenoli is the grave of Pryderi, / Where the waves beat against the land': trans. William F. Skene, 1869, http://www.maryjones.us/ctexts/bbc19.html#4. The Gwenoli is a small stream that flows out of the nearby Llyn Tecwyn Uchaf and joins the Afon Prysor near Ivy Bridge; historically, this would also have been the crossing point over the waters of the estuary, before the floodplain was drained for agriculture – hence Y Felenrhyd, 'the yellow ford'. See also the Royal Commission on the Ancient and Historical Monuments of Wales, 'On the Trail of Pryderi, King of Dyfed', 27 October 2020, https://rcahmw.gov.uk/on-the-trail-of-pryderi-king-of-dyfed/, which mentions both burial sites and Tyddyn Dewin.

9 A good potted history of Nantlleu, with a collection of historical images, can be found at William Stanier, 'Dyffryn Nantlle', 13 April 2021, https://storymaps.arcgis.com/stories/ca7687a56bb343ab9716af16db3f80dc.

10 *The Mabinogion*, trans. Davies, pp. 62–3. As Sioned Davies notes (p. 244): 'The second stanza presents several difficulties, and many interpretations have been offered. The second line

suggests that the oak tree in which Lleu is perched has supernatural qualities – neither the rain nor heat affects it.'

11 *Eryri* has long been thought to mean the 'abode of eagles', although some now say it means 'highlands' or 'to rise': see Steven Morris, 'Yr Wyddfa: Push for Snowdon to Be Known Only by Welsh Name', *Guardian*, 29 April 2021, https://www.theguardian.com/uk-news/2021/apr/29/yr-wyddfa-calls-snowdon-known-only-by-welsh-name. Golden eagles have been all but extinct in Wales since 1850, but a reintroduction plan is now under consideration: see Ben Frampton, 'Snowdonia Golden Eagle Reintroduction Plan Launched', BBC News, 18 August 2020, https://www.bbc.co.uk/news/uk-wales-53806369.

12 Alan Garner, *The Owl Service* (1967), pp. 15 and 156.

13 *The Book of Taliesin*, trans. Lewis and Williams, p. xiii.

14 'The Battle of the Trees', in *The Book of Taliesin*, trans. Lewis and Williams, pp. 54–62.

15 Ibid., p. 54.

16 For more on Wales' *ffridd* habitat, see the RSPB and Natural Resources Wales, 'Ffridd: A Habitat on the Edge', 2014, http://ww2.rspb.org.uk/Images/ffridd_tcm9-384432.pdf.

17 Gwyneth Lewis, 'Wales Millennium Centre', http://www.gwynethlewis.com/biog_millenniumcentre.shtml.

18 Gwynedd Archaeological Trust,

'Historic Landscape Characterisation: Ardudwy – Area 27: Upper Mountain Slopes, Rhinogau', http://www.heneb.co.uk/hlc/ardudwy/ardudwy27.html.

19 Merlin Sheldrake discusses the science behind the 'Wood Wide Web' in his book *Entangled Life: How Fungi Make Our Worlds, Change Our Minds and Shape Our Futures* (2020); see particularly chapter 6.

20 Henry Dimbleby et al., *The National Food Strategy*, 2021, p. 91, https://www.nationalfoodstrategy.org/.

21 Welsh Government, 'Farming Facts and Figures, Wales 2021', 2021, p. 2, https://gov.wales/sites/default/files/statistics-and-research/2021-08/farming-facts-and-figures-2021-695.pdf; Welsh Government, 'Mid-Year Estimates of the Population: 2020', 25 June 2021, https://gov.wales/mid-year-estimates-population-2020.

22 See GetRawMilk.com, 'List of Countries with More Sheep than People: Numbers and Ratios, 2020', 23 January, 2021, https://getrawmilk.com/content/list-of-countries-with-more-sheep-than-people-numbers-and-ratios-2020. In 2021, following a drought, Australia's sheep flock declined to 68 million, making it outnumber the human population by only 2.7 to 1.

23 *The Mabinogion*, trans. Davies, p. 241. Sheep occur only twice in *The Mabinogion*: in 'Peredur

Son of Efrog': 'On one side of the river he could see a flock of white sheep' (p. 89); and in 'How Culhwch Won Olwen': 'they could see a huge flock of sheep without boundary or border to it, and a shepherd on top of a mound tending the sheep' (p. 190). Swine occur in the following passages. 'The Second branch' of the Mabinogi': 'Matholwch's swineherds were on the sea shore one day, busy with their pigs' (p. 29). 'The Fourth Branch': Gwydion stealing Pryderi's pigs (pp. 49–50); the sow that eats the rotting flesh falling off the wounded Lleu Llaw Gyffes, who has changed into an eagle, perched in an oak tree (p. 62). 'Peredur Son of Efrog', describing a feast: 'chops of the flesh of sucking-pigs' (p. 67). 'Lludd and Llefelys', discussing the magical race known as Coraniaid: 'when they are exhausted . . . they will fall onto the sheet in the shape of two little pigs' (p. 113). 'How Culhwch Won Olwen': the opening paragraph of the tale talks about the queen giving birth to a son in a pigpen where a swineherd is tending his herd of pigs (p. 179); Ol son of Olwydd, 'his father's pigs were stolen', but he got them back (p. 188); Arthur's hunt for the supernatural wild boar Twrch Trwyth, with its seven piglets (pp. 208–11).

24 For a potted history of when sheep first came to Britain, see M. L. Ryder, 'The History of Sheep Breeds in Britain', *The Agricultural History Review*, 12:1 (1964), pp. 1–12, https://bahs.org.uk/AGHR/ARTICLES/12n1a1.pdf. For a map of the huge swathes of land acquired and farmed by the Strata Florida abbey, see Strata Florida Trust, 'The History of Strata Florida', https://www.stratafloridatrust.org/history-new.

25 Welsh Government, 'Farming Facts and Figures, Wales 2021', 2021, p. 4, https://gov.wales/sites/default/files/statistics-and-research/2021-08/farming-facts-and-figures-2021-695.pdf.

26 R. S. Thomas, 'A Peasant', from *The Stones of the Fields*, 1946.

27 R. S. Thomas, 'Reservoirs', from *Not That He Brought Flowers*, 1968.

28 R. S. Thomas, 'Reservoirs', from *Not That He Brought Flowers*, 1968.

29 Adam Price MP, 'Wales: The First and Final Colony', annual address to the Institute of Welsh Politics, 16 November 2009, reproduced in full at https://www.walesonline.co.uk/news/wales-news/wales-first-final-colony---2070487.

30 This definition is that used by Rewilding Britain, 'Defining Rewilding', https://www.rewildingbritain.org.uk/explore-rewilding/what-is-rewilding/defining-rewilding.

31 George Monbiot, *Feral: Searching for Enchantment on the Frontiers of Rewilding*, 2013, p. 65 and p. 153.

32 George Monbiot, *Feral: Searching for Enchantment on the Frontiers of Rewilding*, 2013, pp. 176–7.

33 BBC News, 'Farmers' Union of Wales Wants Rewilding Project Scrapped', 31 July 2019, https://www.bbc.co.uk/news/uk-wales-49186349.

34 Lauren Dean, 'Welsh farmers Fight Rewilding plans which Could Take over 10,000 Hectares', *Farmers Guardian*, 7 August 2019, https://www.fginsight.com/news/news/welsh-farmers-fight-rewilding-plans-which-could-take-over-10000-hectares-91108 (paywall); article reproduced in full at https://www.consultationinstitute.org/consultation-news/welsh-farmers-argue-there-was-no-discussion-no-consultation-no-prior-notice-to-rewilding-project/.

35 Rewilding Britain, 'Four in Five Britons Support Rewilding, Poll Finds', 19 January 2022, https://www.rewildingbritain.org.uk/news-and-views/press-releases-and-media-statements/four-in-five-britons-support-rewilding-poll-finds. Full results of the polling, undertaken 13–14 October 2021, are online here: https://docs.cdn.yougov.com/5rzfcgfis7/Rewilding_RewildingInBritain_211014_W.pdf.

36 The figures of twenty-eight farms and 46,000 acres (seventy-two square miles) come from the Elan Valley Trust, who administer the area on behalf of the landowners, Dwr Cymru (Welsh Water): 'About Elan', https://www.elanvalley.org.uk/about.

6. A Perfect Republic of Shepherds

1 William Wordsworth, *Guide to the Lakes* (5th edition, 1835), (ed.) Ernest de Sélincourt, 1960, p. 67, https://archive.org/details/guidetolakesfif00slgoog/mode/2up.

2 William Wordsworth, *Guide to the Lakes* (5th edition, 1835), (ed.) Ernest de Sélincourt, 1960, p. 45, https://archive.org/details/guidetolakesfif00slgoog/mode/2up.

3 Samuel Taylor Coleridge's 'The Nightingale', in his and Wordsworth's *Lyrical Ballads*, 1798, praised the Lake District's 'vernal showers / That gladden the green earth', creating conditions for a 'mossy forest-dell'. John Keats, in his 29 June–2 July 1818 letter to his brother, described the Borrowdale rainforests, recounting how 'The approach to Derwent Water . . . is richly wooded & shut in with rich-toned Mountains': here he is describing the Borrowdale rainforests. See https://keatslettersproject.com/correspondence/john-keats-walks-the-lakes/

4 Wordsworth, *Guide to the Lakes*, pp. 28 and 45.

5 Ibid., pp. 29–30.

6 Ibid., pp. 43 and 52.

7 Robert Gambles, *Words from the Wildwood: Cumbria's Ancient Place-Names*, 2017, pp. 83–5.

8 Ibid., pp. 91–2.

9 National Trust, 'Wild Ennerdale, Cumbria', https://www.nationaltrust.org.uk/features/wild-ennerdale-cumbri: 'Ennerdale . . . means juniper valley in the ancient Norse language.' The National Trust own around 110,000 acres of the Lake District, about a fifth of the National Park's area: figure obtained by the author measuring GIS maps of the Trust's land.

10 Wordsworth, *Guide to the Lakes*, p. 18.

11 George Monbiot, 'The Lake District's World Heritage Site Status Is a Betrayal of the Living World', *Guardian*, 11 July 2017, https://www.theguardian.com/commentisfree/2017/jul/11/lake-district-world-heritage-site-sheep.

12 Lake District National Park Partnership, *Nomination of the English Lake District for Inscription on the World Heritage List*, 2017, chapter 2b, p. 202, https://www.lakedistrict.gov.uk/__data/assets/pdf_file/0023/127553/2.b-History-and-Development-of-The-English-Lake-District.pdf.

13 Wordsworth, *Guide to the Lakes*, p. 57.

14 Ibid., p. 43.

15 Official DEFRA statistics on the number of sheep in the Lake District in the 1950s versus the 2010s are analysed by Miles King in his blog 'Sheepwrecked or a World Heritage Site? Thoughts on the Lake District', A New Nature Blog, 22 May 2017, https://anewnatureblog.com/2017/05/22/sheepwrecked-or-a-world-heritage-site-thoughts-on-the-lake-district/.

16 Wordsworth, *Guide to the Lakes*, p. 52.

17 Lee Schofield, 'Holes in the Map, Part 9: Wolves', Wild Haweswater, 13 March 2021, https://wildhaweswater.co.uk/2021/03/13/holes-in-the-map-part-8-wolves/. Wolf Crags is on Matterdale Common.

18 Wordsworth, *Guide to the Lakes*, p. 61.

19 Thanks to Luke Barley at the National Trust for investigating the question of Johnny Wood's name, and for providing me with this information from colleagues.

20 Met Office Official Blog, 'Did Climate Change Have an Impact on Storm Desmond?', 7 December 2015, https://blog.metoffice.gov.uk/2015/12/07/did-climate-change-have-an-impact-on-storm-desmond/.

21 Natural England, 'Designated Sites View: SSSI Condition Summary – Whole of England', report current as of 27 May 2022, https://designatedsites.naturalengland.org.uk/ReportConditionSummary.aspx?SiteType=ALL

22 Natural England, 'Condition of SSSI Units for Site Johnny Wood SSSI', current as of 27 May 2022: https://designatedsites.naturalengland.org.uk/ReportUnitCondition.aspx?SiteCode=S100028

0&ReportTitle=Johnny%20 Wood%20SSSI.

23 National Trust Land Map: entries for Johnny Wood show part of it was bought by the Trust in 1959, and the remainder in 1964. See https:// www.nationaltrust.org.uk/ features/follow-the-history-of-our-places-with-land-map.

24 Lee Schofield, *Wild Fell: Fighting for Nature on a Lake District Farm*, 2022, pp. 16–17.

25 The Haweswater vision images are shown in this video by RSPB and United Utilities, 'A Vision for Haweswater', 23 February 2022: https://www. youtube.com/ watch?v=zjpOAIS6EPE. On eagle reintroduction, see also Lee Schofield, 'Holes in the Map, Part 4: Eagles', Wild Haweswater, 14 November 2020, https://wildhaweswater. co.uk/2020/11/14/holes-in-the-map-part-3-eagles/.

26 RSPB and United Utilities, 'Securing Nature, Serving People: The Next 40 Years at Haweswater', 2020, https:// wildhaweswater.co.uk/ wp-content/uploads/ Haweswater-Securing-Nature-Serving-People.pdf.

27 David Harvey and Charles Scott, 'Farm Business Survey 2015/16: A Summary from Hill Farming in England', Newcastle University Rural Business Research report, March 2017, pp. iii–iv. See also Duchy College Rural Business School, 'The Value of the Sheep Industry: North East, South West and North West Regions', February 2018, p. 17, https:// www.nfuonline.com/ archive?treeid=106083.

28 See, for example, Brenda Mayle, 'Domestic Stock Grazing to Enhance Woodland Biodiversity', Forestry Commission information note, September 1999, p. 3, http:// adlib.everysite.co.uk/ resources/000/111/078/fcin28. pdf, which states that sheep 'browse hardwood seedlings for a greater proportion of [the] year and more selectively than cattle, preventing natural regeneration'.

29 Chris Clark, Brian Scanlon and Kaley Hart, 'Less Is More: Improving Profitability and the Natural Environment in Hill and Other Marginal Farming Systems', report for the Wildlife Trusts, RSPB and the National Trust, November 2019, https:// www.wildlifetrusts.org/sites/ default/files/2019-11/Hill%20 farm%20profitability%20 report%20-%20FINAL%20 agreed%2015%20Nov% 2019.pdf.

30 United Utilities, 'Cryptosporidium', 2016, https://www.unitedutilities.com/ globalassets/documents/pdf/ cryptosporidium_acc16.pdf: 'A very effective way to minimise the risk from Cryptosporidium is to protect the raw water sources from contamination by careful catchment management.'

31 Lee also discusses bracken in his book. 'Bracken only thrives

where the soils are deep and dry enough for trees to grow . . . Farmers hate bracken, walkers hate bracken . . . the obvious thing is to plant trees into it.' Schofield, *Wild Fell*, pp. 161–3.

32 Open Spaces Society, 'Commons: Buildings, Fences and Other Works on Common Land in England', last revised July 2021, https://www.oss.org. uk/need-to-know-more/ information-hub/buildings-fences-and-other-works-on-common-land-in-england/.

33 See Nofence's website: https:// www.nofence.no/en/.

34 John Evelyn, *Sylva*, 1664.

35 Lord Bolton, *Profitable Forestry*, 1956, p. 96.

36 Ralph Harmer and Gary Kerr, 'Natural Regeneration of Broadleaved Trees', Forestry Commission, Research information note, October 1995, https://www. forestresearch.gov.uk/ documents/997/rin275.pdf.

37 John Nisbet, *Studies in Forestry: Being a Short Course of Lectures on the Principles of Sylviculture*, 1894, quoted in Esmond Harris, Jeanette Harris and N. D. G. James, *Oak: A British History*, 2003, p. 33.

38 Peter S. Savill, *The Silviculture of Trees Used in British Forestry*, 1991, quoted in Harris et al., *Oak*, p. 35.

39 Bolton, *Profitable Forestry*, p. 98.

40 These factors and others are discussed in detail in Harris et al., *Oak*, pp. 32–5, and by R. Worrell and C. J. Dixon, 'Factors Affecting the Natural Regeneration of Oak in Upland Britain: A Literature Review', Forestry Commission occasional paper 31, 1991, https://www. forestresearch.gov.uk/ documents/6874/FCOP031. pdf.

41 Worrell and Dixon, 'Factors Affecting the Natural Regeneration'.

42 Isabella Tree, *Wilding: The Return of Nature to a British Farm*, 2018, p. 124.

43 Arthur Standish, *New Directions of Experience to the Commons Complaint*, 1613, p. 9, https:// quod.lib.umich.edu/e/eebo2/ B08132.0001.001?rgn=m ain;view=fulltext. Many thanks to Philip Amies for locating this original reference.

44 'Word of the day' tweet by Robert Macfarlane, 12 October 2018, https://twitter.com/ robgmacfarlane/ status/1050627236934516736? lang=en-GB and India Bourke, 'Where have all the jays come from?', *New Statesman,* 19 April 2022, https://www. newstatesman.com/culture/ nature/2022/04/where-have-all-the-jays-come-from.

45 Patrick Barkham, 'Half the Trees in Two New English Woodlands Planted by Jays, Study Finds', *Guardian*, 16 June 2021, https://www. theguardian.com/ environment/2021/jun/16/half-the-trees-in-two-new-english-woodlands-planted-by-jays-study-finds. The original study

is Richard K. Broughton et al., 'Long-Term Woodland Restoration on Lowland Farmland through Passive Rewilding', *PLoS ONE*, 16:6, 2021, e0252466, https://doi.org/10.1371/journal.pone.0252466.

46 Clifton Bain, *The Rainforests of Britain and Ireland: A Traveller's Guide*, 2015, p. 28.

47 C. D. Pigott, 'Regeneration of Oak–Birch Woodland Following Exclusion of Sheep', *Journal of Ecology*, 71:2 (1983), p. 643, https://www.jstor.org/stable/2259738?origin=crossref&seq=1. I'm grateful to Luke Barley at the National Trust for making me aware of this study and for sending me a copy. The ecologist Ian Rotherham also visited Yarncliff Wood in 2020 and reported on how the woodland behind the exclosure fencing continues to thrive: see Ian Rotherham and Ondřej Vild, 'Long-Term Exclosure of Sheep-Grazing from an Ancient Wood: Vegetation Change after a Sixty-Year Experiment', International Association of Vegetation Science, 7 December 2020, https://vegsciblog.org/2020/12/07/long-term-exclosure-sheep-grazing-ancient-wood/.

48 A. G. Tansley, *Britain's Green Mantle: Past, Present and Future*, 1949, pp. 111–12.

49 Maurice Pankhurst, 'Keskadale – An Oakwood on the Edge', *British Wildlife*, 19:3 (2008), pp. 187–92, https://www.britishwildlife.com/article/volume-19-number-3-page-187-192.

50 See the Piles Copse website: https://pilescopse.org.uk/.

51 Obituary of Henry Hurrell, Devonshire Association, 1981, https://devonassoc.org.uk/person/hurrell-h-g/.

52 Linda Geddes, 'Britons Cut Meat-Eating by 17%, but Must Double that to Hit Target', *Guardian*, 8 October 2021, https://www.theguardian.com/food/2021/oct/08/cuts-uk-meat-consumption-doubled-health-researchers-food.

53 James Rebanks, *English Pastoral: An Inheritance*, 2020, p. 272.

54 April Windle, 'Riparian Woodlands and Woody Debris Creation: Considerations for Lichens and Bryophytes', *Conservation Land Management*, 18:3 (2020), pp. 15–22.

55 Derek Ratcliffe, 'An Ecological Account of Atlantic Bryophytes in the British Isles', *New Phytologist*, 67 (1968), pp. 365–439. The woodland ecologist George Peterken also states that 'oceanic bryophytes . . . are sensitive to both edges and canopy removal': 'Reversing the Habitat Fragmentation of British Woodlands', WWF-UK report, January 2002, p. 30, https://wwfeu.awsassets.panda.org/downloads/ukforestsfragmentation.pdf.

56 Kent Wildlife Trust, 'Wilder Blean', https://www.kentwildlifetrust.org.uk/wilderblean, and Damian Carrington, '"Gentle Giants":

Rangers Prepare for Return of Wild Bison to UK', *Guardian*, 11 December 2021, https://www.theguardian.com/environment/2021/dec/11/gentle-giants-rangers-prepare-return-wild-bison-uk.

57 April Windle, 'Riparian Woodlands and Woody Debris Creation'.

58 Vincent Wildlife Trust, 'Pine Marten', 12 March 2015, https://www.vwt.org.uk/species/pine-marten/, and Forestry England, 'The Return of Pine Martens to England's Forests', https://www.forestryengland.uk/blog/the-return-pine-martens-englands-forests. Studies show that pine martens eat many more grey squirrels than they do reds, because red squirrels have co-evolved with pine martens and so know better how to evade them. See Joshua P. Twining, W. Ian Montgomery and David G. Tosh, 'The Dynamics of Pine Marten Predation on Red and Grey Squirrels', *Mammalian Biology*, 100 (2020), pp. 285–93, https://link.springer.com/article/10.1007%2Fs42991-020-00031-z.

59 See, for example, Michel Meuret and Elodie Regner, 'Wolves and Livestock Farming in France: An Assessment of 27 years of Coexistence', French National Research Institute for Agriculture, Food and Environment (INRAE), 9 April 2020, https://www.inrae.fr/en/news/wolves-and-livestock-farming-france-assessment-27-years-coexistence, and WWF, 'Into the Wilderness of Bureaucracy: Coexistence with Large Carnivores in Romania', 2 April 2019, https://wwf.panda.org/wwf_news/?345094/Into-the-Wilderness-of-Bureaucracy.

7. 'Nobody Cares for the Woods Anymore'

1 Dick Jude, *Fantasy Art Masters: The Best Fantasy and SF Art Worldwide*, 1999, p. 26.

2 Catherine Courtenay, 'Dartmoor's Oscar-Winning Artist Alan Lee on Why He Loves the Area', *Great British Life*, 1 May 2020, https://www.greatbritishlife.co.uk/people/dartmoor-s-oscar-winning-artist-alan-lee-on-why-he-7276484.

3 Jude, *Fantasy Art Masters*, p. 18.

4 See Alan Lee, *The Entwash* (illustration for Book III, Chapter III, 'The Uruk-hai'), and *Searching Fangorn* (illustration for Book III, Chapter V, 'The White Rider'), in J. R. R. Tolkien, *The Lord of the Rings: The Two Towers*, illustrated edition, 1991.

5 J. R. R. Tolkien, *The Lord of the Rings*, 1973 (single volume edition), p. 482.

6 Ibid., p. 484.

7 Laura Joint, 'Trees as Art', BBC Devon, 5 October 2007, http://www.bbc.co.uk/devon/content/articles/2007/10/04/trees_alan_lee_feature.shtml.

8 The Devonian local historian Francis Billinge examined the evidence for whether Tolkien

ever visited Dartmoor. He concluded that there was none, but that Tolkien did have a relative who lived in Haytor Vale where he could have once stayed: 'Tolkien and Haytor Vale', Aspects of the History of Bovey Tracey, 2020, https://boveytraceyhistory.org.uk/479-2/suez-canal-links-with-bovey-tracey/.

9 Quoted in Michael C. Drout (ed.), *J. R. R. Tolkien Encyclopedia: Scholarship and Critical Assessment*, 2006, p. 678, under the entry for 'Trees'.

10 Meredith Veldman, *Fantasy, the Bomb, and the Greening of Britain: Romantic Protest, 1945–1980*, 1994, p. 90. Other commentators who have interpreted *The Lord of the Rings* as a reaction against the Industrial Revolution and modern warfare include the historian Dominic Sandbrook, 'This Is Tolkien's World', *Unherd*, 10 December 2021, https://unherd.com/2021/12/this-is-tolkiens-world/; and Professor Richard Gunderman, 'Tolkien and the Machine', *The Conversation*, 6 January 2015, https://theconversation.com/tolkien-and-the-machine-35826.

11 Veldman, *Fantasy, the Bomb, and the Greening of Britain*, p. 109.

12 Dorothy Elmhirst, quoted in Michael Young, *The Elmhirsts of Dartington: The Creation of an Utopian Community*, 1982, p. 4.

13 Young, *The Elmhirsts of Dartington*, pp. 105–6.

14 Ibid., pp. 143 and 247, and *passim*.

15 Map of Britain's oceanic climate zone constructed by the author using hygrothermic grid data supplied by Dr Christopher Ellis; ancient woodland inventory maps sourced from Natural England.

16 Wilfred Hiley, *A Forestry Venture*, 1964, pp. 60 and 67.

17 Dartmoor National Park Authority woodland survey data, shared with the author; with thanks to Richard Knott, ecologist at DNPA. For Fingle Woods and other woods in the Teign valley, the original woodland type is recorded simply as 'WO': Richard Knott tells me that 'this code is used where it was not possible to identify a ground flora vegetation community under mature stands of conifer, because of the heavy shade which suppressed the ground flora'.

18 Edmund Spenser, *The Faerie Queene*, 1590, book I, canto I, verse VIII.

19 John Evelyn, *Sylva, or, A Digression of Forest-Trees and the Propagation of Timber*, 1664, p. 8.

20 Hiley, *A Forestry Venture*, p. 67.

21 H. L. Edlin, *Trees, Woods and Man* (Collins New Naturalist Series), 1956, plate facing p. 212.

22 HMSO, *Dartmoor: National Park Guides Number One*, 1957, pp. 13–14.

23 Lord Bolton, *Profitable Forestry*, 1956, p. 45.

24 Forestry England, 'Timber and the Uses of Wood', https://www.forestryengland.uk/timber-uses-of-wood: 'Softwoods come from coniferous trees such as pine, fir, spruce and larch. These trees take around 40 years to grow before they are ready to harvest. Hardwoods come from broadleaved trees such as oak, ash and beech. These trees take much longer to grow, up to 150 years before they are ready to harvest.'

25 'Dartington Woodlands Ltd: Report for the Eight Months January 1st 1947 to August 31st 1947', Dartington Archives, C/DWL/1/F, – Reports and Accounts, 1947–57.

26 Dartington Archives, C/DWL/21/K, bound maps of Kingswood (1938) and two copies of a proposal for Kingswood.

27 Tolkien, *The Lord of the Rings*, 1973 (single volume edition), p. 494.

28 W. E. Hiley, 'News of the Day: Changes in the Woodland Department', undated (probably 1946), Dartington Archives, C/DWL/1/A, Formation of Company (1947).

29 Hiley, *A Forestry Venture*, p. 69.

30 Letter from Wilfred Hiley to Viscount Newport, 1 June 1949, Dartington Archives, C/DWL/2/E, Siegfried Marian.

31 Young, *The Elmhirsts of Dartington*, pp. 288–9.

32 Edlin, *Trees, Woods and Man*, p. 52.

33 T. R. E. Southwood, 'The Number of Species of Insect Associated with various Trees', *Journal of Animal Ecology*, 30:1 (1961), p. 1.

34 Derek Ratcliffe, 'An Ecological Account of Atlantic Bryophytes in the British Isles', *New Phytologist*, 67 (1968), pp. 365–439.

35 Wilfred Hiley, draft synopsis for *A Forestry Venture*, February 1959, Dartington Archives, T/HIL/3/E, correspondence re. Hiley's book 'A Forestry Venture'.

36 Michael Young, *The Elmhirsts of Dartington*, p. 287. See also Victor Bonham-Carter, *Dartington Hall: The History of an Experiment*, 1958, p. 153.

37 Forestry Commission letter to Dartington Woodlands Ltd 16 June 1954, Dartington Archives, C/DWL/19/Q, Scrub Clearance Scheme, 1954–9.

38 Oliver Rackham, *Trees and Woodland in the British Landscape*, 1976, p. 174.

39 As Rackham says (*Trees and Woodland*, p. 173): 'Evidence of destruction is all around us, but to measure it would require a considerable research programme. Successive Forestry Commission censuses were not compiled for this purpose and are of little use to us . . . ancient woodland forms an unknown fraction of various categories from Broadleaved High Forest to Unutilizable Scrub.'

40 Ancient Woodland Inventory for England published by

Natural England: https://naturalengland-defra.opendata.arcgis.com/datasets/Defra::ancient-woodland-england/about. Ancient Woodland Inventory for Wales published by Natural Resources Wales: http://lle.gov.wales/catalogue/item/AncientWoodlandInventory2021/?lang=en. Ancient Woodland Inventory for Scotland published by Scottish Natural Heritage: https://www.nature.scot/doc/guide-understanding-scottish-ancient-woodland-inventory-awi.

41 Peter Marren, *Nature Conservation: A Review of the Conservation of Wildlife in Britain, 1950–2001* (Collins New Naturalist series), 2002, p. 160.

42 Woodland Trust, 'Our Founder – Kenneth Watkins OBE (1909–1996)', https://www.woodlandtrust.org.uk/about-us/working-with-us/our-people/our-founder/, states that 'Ken's ashes are scattered in King's Wood – his favourite sitting place'. King's Wood (not to be confused with the former Dartington woodland also mentioned in this chapter) lies within Watkins's Hall Farm Estate. The Woodland Trust describes it as 'Atlantic oak woodland' and 'abundant with ferns, mosses and lichens': 'Hall Farm Estate Management Plan, 2018–2023', pp. 9 and 15, https://www.woodlandtrust.org.uk/media/46788/4178-hall-farm-estate.pdf.

43 I'm grateful to Kevin Cox, chair of the RSPB, for recounting this story to me, and to Leonard Hurrell, Henry Hurrell's son, for allowing me to visit his home and woods where the Woodland Trust was founded. For more on the early history of the Woodland Trust, see Marren, *Nature Conservation*, pp. 74 and 172, and David Evans, *A History of Nature Conservation in Britain*, 1997 edition, p. 126.

44 Woodland Trust, 'Avon Valley Woods Management Plan, 2017–2022', https://www.woodlandtrust.org.uk/media/46633/4001-avon-valley-woods.pdf: 'The very first wood owned by the Trust was Bedlime which was purchased on the 24th October 1972. At that time, the wood had been used widely for shooting and there was a perceived pressure that woods within the valley would be felled and converted to conifer as part of the drive by the government policies and management fashions of the time' (p. 6). The plan describes Bedlime Wood as 'ancient semi natural woodland, upland oak with close affinity to NVC type W10' (p. 22).

45 Dominic Sandbrook, *State of Emergency: The Way We Were: Britain, 1970–1974*, 2010, p. 200.

46 Woodland Trust, 'Fingle Woods', https://www.woodlandtrust.org.uk/visiting-woods/woods/fingle-woods/.

47 Miles King and John Underhill-Day, 'Fingle Woods Conservation Plan – Information to Inform HLF Bid', Footprint Ecology, 2015, p. 48, https://www.footprint-ecology.co.uk/reports/Underhill-Day%20and%20King%20M.%20-%20 2015%20-%20Fingle%20 Woods%20conservation%20 plan%20-%20information%20 to%20in.pdf.

48 Ibid., pp. 33 and 39.

49 The British Trust for Ornithology records that British populations of pied flycatchers fell during the 1990s, stabilised between 2005 and 2015, but have then started to decline again: 'Bird Trends: Pied Flycatcher', https://app.bto.org/birdtrends/species.jsp?year=2019&s=piefl.

50 Woodland Trust, 'Appeals: Ausewell Wood', https://www.woodlandtrust.org.uk/support-us/give/appeals/ausewell-wood/, and Colleen Smith, 'Devon's "Lost World" Rainforest Saved after Amazing Public Appeal', *Devon Live*, 18 February 2020, https://www.devonlive.com/news/devon-news/devons-lost-world-rainforest-saved-3860290.

51 In 1985, the Forestry Commission introduced the 'Broadleaves Policy' and a new set of grants to support the planting of native broadleaved trees: see Forestry Commission, 66th annual report and accounts, 1985–6. In 2007, Forestry England made a commitment to begin restoring the 43,000 hectares of PAWS in its possession – although details of that commitment appear to have disappeared from the internet, and progress remains slow. Thanks to Nick Phillips and Eleanor Lewis at the Woodland Trust for looking into this for me.

52 There are 135,000 hectares of PAWS in England; with 43,000 hectares of these owned by Forestry England, that leaves around 92,000 hectares in private ownership. See Woodland Trust, 'The Current State of Ancient Woodland Restoration', January 2018, p. 7, https://www.woodlandtrust.org.uk/media/1704/current-state-of-ancient-woodland-restoration.pdf.

53 Forestry Commission, 'Key Performance Indicators: Report for 2020–21', June 2021, p. 50, https://assets.publishing.service.gov.uk/government/uploads/system/uploads/attachment_data/file/1002042/Forestry-Commission-Key-Performance-Indicators-Report-for-2020-21-.pdf.

54 Quoted in 'The Dartmoor Society 13th Annual Debate: What Future for Dartmoor Woodlands?', 23 October 2010, https://www.dartmoorsociety.com/debates/2010.

55 You can see the tweet here: https://twitter.com/guyshrubsole/status/14454613 40747829249.

8. The Accidental Rainforest

1 Lustleigh Parish Council, 'A Guide to the Parish of Lustleigh in the County of Devon', 2019. It was once thought that the Anglo-Saxon name 'Sutreworde' meant 'South of the Wood', but more recent research by local historian Ian Mortimer has suggested it in fact means 'south enclosure'. Lustleigh Society, 'The History of Lustleigh' https://www.lustleigh-society.org.uk/home-page/history/. Nevertheless, the Domesday Book records Sutreworde as having woodland spanning 1 league by half a league, a considerable area; see Open Domesday project entry for Sutreworde, https://opendomesday.org/place/XX0000/sutreworde/.

2 For a reproduction of Francis Stevens's painting *Lustleigh Cleave*, see Peter Mason, *Dartmoor, a Wild and Wondrous Region: The Portrayal of Dartmoor in Art, 1750–1920*, 2017, p. 28 (exhibition catalogue for a show that ran at the Royal Albert Memorial Museum in Exeter in 2017–18). The painting can also be viewed online at https://www.peterstephens.co.uk/j-f-stevens-painting/.

3 Samuel Rowe, *A Perambulation of Dartmoor*, 1848, pp. 136–7. Rowe undertook his fieldwork in 1827–8, but the book was not published until later.

4 For the first edition Ordnance Survey map of Lustleigh Cleave (surveyed 1803–8), see the records of the Charles Close Society: https://www.charlesclosesociety.org/files/Oldseries/OldseriesIndex.htm – click on Sheet 25 (Tavistock). For the second edition Ordnance Survey maps, see the National Library of Scotland's 'OS Sheets Record Viewer', https://maps.nls.uk/: load up the 6 Inch series maps from 1888–1913 or the 25 Inch series from 1873–1888. All of these maps show Lustleigh Cleave as remaining almost entirely bare of trees during the nineteenth century.

5 William Crossing, *Guide to Dartmoor*, 1909, p. 291.

6 The RAF's 1946 aerial photographs are published on Devon County Council's Environment Viewer map, at https://maptest.devon.gov.uk/portaldvl/apps/webappviewer/index.html?id=82d17ce243be4ab28091ae1f15970924 – select 'Aerial RAF 1946' from the menu of basemap layers.

7 'Panorama near Manaton, looking NE', 1962, Cambridge University Department of Geography, Cambridge Air Photos, https://www.cambridgeairphotos.com/location/afg91/.

8 David & Charles (Publishers) Ltd, *Dartmoor in Pictures*, 1971, p. 1. The author holds a copy of this pamphlet.

9 T. R. Harrod, 'Soils in Devon IX', Soil Survey Record No. 117, 2017, p. 343, https://www.cranfield.ac.uk/-/media/

files/landis-downloads/r117-soils-in-devon-ix.ashx.

10 Google Earth aerial imagery for 2001 and 2021 shows vegetation continuing to creep up Lustleigh Cleave.

11 Lustleigh Society, 'The Ruins of Boveycombe', December 2015 (extract from the Lustleigh parish magazine, November 2011), https://www.lustleigh-society.org.uk/the-ruins-of-boveycombe/.

12 Esther Addley, 'Ben-Fur: Romans Brought Rabbits to Britain, Experts Discover', *Guardian*, 18 April 2019, https://www.theguardian.com/lifeandstyle/2019/apr/18/ben-fur-romans-brought-rabbits-to-britain-experts-discover. Large-scale breeding of rabbits in Britain, however, only really started after the Norman Conquest in 1066.

13 John Martin, 'The Wild Rabbit: Plague, Policies and Pestilence in England and Wales, 1931–1955', *The Agricultural History Review*, 58:2 (2010), p. 275, https://www.bahs.org.uk/AGHR/ARTICLES/58_2_6_Martin.pdf.

14 Lustleigh Society Archives, BOX B/007 (folder section on vegetation management on the Cleave): parish newsletter article by Jill Salmon, clerk of the Lustleigh Commoners Association, July 1985.

15 Around the year 1600, some 27–30 per cent of England was common land, according to Gregory Clark and Anthony Clark, 'Common Rights to Land in England, 1475–1839', *The Journal of Economic History* 61:4 (2001), pp. 1009–36. Nowadays, common land covers just 3 per cent of England, around 1 million acres: see Foundation for Common Land, 'What Is Common Land?', https://foundationforcommonland.org.uk/a-guide-to-common-land-and-commoning.

16 Commons in Wales cover some 295,000 acres, a figure obtained by the author by measuring the maps of Welsh common land published by Natural Resources Wales (see http://lle.gov.wales/catalogue/item/OpenAccessRegisteredCommonLand). Wales is some 5 million acres, so commons comprise around 6 per cent of the country. In Scotland, most Lowland commons were enclosed, but crofting common grazings still cover about 1.3 million acres in the Highlands and Islands, or about 7 per cent of Scotland's total area of 19 million acres: see Andy Wightman et al., 'Common Land in Scotland: A Brief Overview', December 2003, p. 9, http://www.andywightman.com/docs/secur_comm8.pdf.

17 Garrett Hardin, 'The Tragedy of the Commons', *Science*, 162, no. 3859 (13 December 1968), pp. 1243–8, https://pages.mtu.edu/~asmayer/rural_sustain/governance/Hardin%201968.pdf.

18 A good summary of the arguments of Ostrom and

Hardin can be found in Matthew Kelly, *Quartz and Feldspar – Dartmoor: A British Landscape in Modern Times*, 2015, pp. 357–8.

19 The Verderers of the New Forest were put on a statutory footing under the New Forest Act of 1887, but had existed as a decision-making body since the thirteenth century. See https://www.verderers.org.uk/.

20 Peter Linebaugh, *The Magna Carta Manifesto: Liberties and Commons for All*, 2008, p. 9 ('mutuality and negotiation'); George Monbiot, 'Common Wealth', 2 October 2017 ('destroys inequality' and 'provides and incentive . . .'), https://www.monbiot. com/2017/10/02/common-wealth/. The economist Guy Standing's book *The Plunder of the Commons: A Manifesto for Sharing Public Wealth*, 2019, is another good overview of the enclosure of the commons and a manifesto for reclaiming the concept of 'commons' for the twenty-first century. For a further left–green defence of commons, see Simon Fairlie, 'A Short History of Enclosure in Britain', *The Land*, 7 (2009), https://www.thelandmagazine. org.uk/articles/short-history-enclosure-britain.

21 Figures for the number of cows, sheep and ponies allowed to graze on Lustleigh Common all come from the Lustleigh Society Archives, BOX B/005: document listing all the registered commoners (undated,

but it appears to be from the 1990s or 2000s, as it also lists email addresses). A separate list of registered commoners and their rights is also published by Devon County Council: see https://maptest.devon.gov.uk/ arcgisdvl/rest/services/ Highways_Intranet/ CommonLandVillageGreens/ MapServer/0/163/ attachments/1674.

22 A. R. Sibbald et al., 'Heather Moorland Management: A Model', in M. Bell and R. G. H. Bunce (eds), *Agriculture and Conservation in the Hills and Uplands*, 1987, pp. 107–8. I am grateful to Adrian Colston for drawing my attention to this study in his PhD thesis, 'Stakeholder narratives of Dartmoor's Commons: tradition and the search for consensus in a time of change', April 2021, https:// adriancolston.files.wordpress. com/2021/07/thesis-full-v5-viva-corrections.pdf.

23 See entry in The Peerage for Lt.-Col. Fleetwood Hugo Pellow, http://www.thepeerage. com/p40222.htm

24 See entries in The Peerage for Lt.-Col. John Francis Whidborne, http://www. thepeerage.com/p56996. htm#i569954, and for Capt. Sandro Ansell George Stuart Bullock-Webster, https://www. thepeerage.com/p42080. htm#i420791.

25 See entry in The Peerage for Lady Susan Jean Carnegie (whose married name became

Connell), http://www.
thepeerage.com/p4801.
htm#i48010. Records of her
common grazing rights come
from the Lustleigh Society
Archives, BOX B/005:
document listing all the
registered commoners (undated,
but appears to be from 1990s–
2000s).

26 Records of James Paxman's
common grazing rights come
from the Lustleigh Society
Archives, BOX B/005:
document listing all the
registered commoners (undated,
but appears to be from 1990s–
2000s).

27 Lustleigh Society Archives,
BOX B/007: 'Lustleigh Cleave:
The Commoners Management
Objectives', June 1997.

28 Ibid.

29 Lustleigh Society Archives,
BOX B/006: letter from Miss J.
Salmon to the editor of the
Western Morning News, 29
October 1976.

30 Lustleigh Society Archives,
BOX B/005: minutes of the
annual general meeting of the
Lustleigh Cleave Commoners
Association, 6 June 1995.

31 Lustleigh Society Archives,
BOX B/005: minutes of
Lustleigh Commoners
Association meeting, 12
February 1993.

32 Lustleigh Society Archives,
BOX B/007: correspondence
between the Nature
Conservancy Council's senior
warden, David Rogers, and
Miss J. Salmon, clerk of the
Lustleigh Commoners

Association, 26 November
1981.

33 Commons Commissioner's
ruling 'In the Matter of
Lustleigh Cleave, Lustleigh', 27
January 1983, available online
at https://www.acraew.org.uk/
sites/default/files/uploads/
Devon/LUSTLEIGH%20
CLEAVE%20-%20
LUSTLEIGH%20NO.
CL.57(2).pdf.

34 Lustleigh Society Archives,
BOX B/007: 'Lustleigh Cleave:
The Commoners Management
Objectives', June 1997 (bracken
crushing and clearing parties);
BOX B/007, letter from Mrs
J. M. K. Hewison of the
Lustleigh Commoners
Association to Dr R. J. Wolton
of English Nature (cc'd to
Dartmoor National Park
Authority), 23 February 1993
(reinstatement of swaling);
BOX B/007: cuttings from
Lustleigh parish magazine
recounting clearing parties and
swaling events, August 1993–
March 1994.

35 Lustleigh Society Archives,
BOX B/005: report by
Professor Roy Brown, 'Lustleigh
Cleave (CL57) Management
Proposals', May 1999. For a
contemporary newspaper article
about the Bracken Advisory
Commission and Professor
Brown's role in it, see Geoffrey
Lean, 'It's Green, It's Pretty,
and It Can Kill You',
Independent, 3 August 1996,
https://www.independent.co.uk/
news/uk/home-news/it-s-green-
it-s-pretty-and-it-can-

kill-you-1308094.html.

36 Lustleigh Society Archives, BOX B/007: newspaper cutting, 'Bid to Halt Bracken March', *Westcountry News*, 7 October 1996. One medical review of the health impacts of bracken has concluded 'there is no convincing evidence that people are at serious risk from it': Deborah Wilson, Liam J. Donaldson and Ovnair Sepai, 'Should We Be Frightened of Bracken? A Review of the Evidence', *Journal of Epidemiology and Community Health*, 52 (1998), p. 816, online at https://jech.bmj.com/content/jech/52/12/812.full.pdf.

37 The recommendation of aerial spraying with a chemical compound is referred to in Lustleigh Society Archives, BOX B/007: newspaper cutting, 'Bid to Halt Bracken March', *Westcountry News*, 7 October 1996. Use of Asulam was discussed by the Lustleigh commoners in 1995: see Lustleigh Society Archives, BOX B/007: letter from local residents Peter and Nina van Moorsel to Mrs J. Hewison, chair of the Lustleigh Commoners Association, June 22 1995, raising their concerns about the use of Asulam, enclosing a leaflet from the Pesticides Trust and proposing alternative measures.

38 Lustleigh Society Archives, BOX B/007: correspondence between the Nature Conservancy Council's senior warden, David Rogers, and Miss J. Salmon, clerk of the Lustleigh Commoners Association, 26 November 1981.

39 Lustleigh Society Archives, BOX B/007, letter from Mrs J. M. K. Hewison of the Lustleigh Commoners Association to Dr R. J. Wolton of English Nature (cc'd to Dartmoor National Park Authority), 23 February 1993.

40 Lustleigh Society Archives, BOX B/005: minutes of a general meeting of Lustleigh Cleave Commoners Association held at South Harton, 17 September 1996. The minutes record the commoners unanimously agreed to join the Environmentally Sensitive Areas scheme, and that it would last for ten years.

41 See the map compiled by Anna Powell-Smith, 'Farm Payments for Environmental Stewardship', March 2017, https://farmpayments.anna.ps/. The Environmental Stewardship scheme replaced the old Environmentally Sensitive Areas programme; the agreement for Lustleigh Cleave looks to have spanned the ten years from 2008 to 2017 and totalled £216,594.

42 Oliver Rackham, *Woodlands* (Collins New Naturalist Series), 2006.

43 Kelly, *Quartz and Feldspar*, p. 423.

44 The common in question is Brent Moor. The result of the court case referred to, Dance v. Savery & Ors (2011), can be

found at https://www.casemine.
com/judgement/
uk/5a8ff70d60d03e7f57ea6c67.

45 For an account of this progress,
 see two posts by Adrian
Colston on his Dartmoor Blog,
written five years apart: 'A Trip
to Holne Moor – Cuckoos,
Scrub and Flood Prevention', 9
February 2016, https://
adriancolston.wordpress.
com/2016/02/09/a-trip-to-
holne-moor-cuckoos-scrub-and-
flood-prevention/, and
'Re-Wetting and Slowing the
Flow on Holne Moor', 30 May
2021, https://adriancolston.
wordpress.com/2021/05/30/
re-wetting-and-slowing-the-
flow-on-holne-moor/.

46 Phoebe Weston,
'"Un-Managing the Land":
Sheep Make Way for Trees in
Cumbria's Uplands', *Guardian*,
30 October 2020, https://www.
theguardian.com/
environment/2020/oct/30/
un-managing-the-land-the-hill-
farmers-helping-to-rewild-
britain-aoe.

9. Forest People

1 Wikipedia entry for Narubia
Werreria, https://pt-m-
wikipedia-org.translate.goog/
wiki/Narubia_Werreria?_x_tr_
sl=pt&_x_tr_tl=en&_x_tr_
hl=en&_x_tr_pto=sc.

2 Val Munduruku, 'How I Took
the Munduruku Fight to Save
the Amazon to the World
Stage', OpenDemocracy, 22
April 2021, https://www.
opendemocracy.net/en/
democraciaabierta/munduruku-

fight-save-amazon-indigenous-
world-deforestation/.

3 Phoebe Weston, 'Indigenous
Peoples to Get $1.7bn in
Recognition of Role in
Protecting Forests', *Guardian*, 1
November 2021, https://www.
theguardian.com/
environment/2021/nov/01/
cop26-indigenous-peoples-to-
get-17bn-in-recognition-of-role-
in-protecting-forests-aoe.

4 'IUCN Director General's
Statement on International Day
of the World's Indigenous
Peoples', International Union
for Conservation of Nature, 9
August 2019, https://www.iucn.
org/news/secretariat/201908/
iucn-director-generals-
statement-international-day-
worlds-indigenous-
peoples-2019.

5 Food and Agriculture
Organization of the United
Nations, 'New Report Shows
Indigenous and Tribal Peoples
"Best Guardians" of Forests', 25
March 2021, https://www.fao.
org/newsroom/detail/
New-report-shows-Indigenous-
and-Tribal-Peoples-best-
guardians'-of-forests/en. The
report found that 'Almost half
(45 percent) of the intact
forests in the Amazon Basin are
in indigenous territories', and
that 'Brazil's indigenous
territories have more species of
mammals, birds, reptiles, and
amphibians than in all the
country's protected areas
outside these territories'.

6 All Bolsonaro quotations in this
paragraph are cited, with

sources, by Survival International, 'What Brazil's President, Jair Bolsonaro, Has Said about Brazil's Indigenous Peoples', https://www.survivalinternational.org/articles/3540-Bolsonaro.

7 Reuters, 'Deforestation in Brazil's Amazon at Highest Level since 2006', *Guardian*, 19 November 2021, https://www.theguardian.com/environment/2021/nov/18/deforestation-in-brazils-amazon-rises-by-more-than-a-fifth-in-a-year.

8 Scottish government, 'Languages', https://www.gov.scot/policies/languages/gaelic/: 'As the custodian of Scottish Gaelic we have a duty to protect this indigenous language.'

9 On his Twitter profile, Àdhamh Ó Broin describes himself as an 'indigenous activist': https://twitter.com/gaeliconsultant.

10 Alastair McIntosh, *Soil and Soul: People versus Corporate Power*, 2001, p. 19; the whole of Part One is entitled 'Indigenous Childhood; Colonial World'.

11 Trade, Development and the Environment Hub, 'Novel Study Maps out the Inequality of Land Distribution and Ownership in Brazil', 10 August 2020, https://tradehub.earth/2020/08/10/novel-study-maps-out-the-inequality-of-land-distribution-and-ownership-in-brazil/.

12 Andy Wightman, *The Poor Had No Lawyers: Who Owns Scotland (And How They Got It)*, 2015 edition, p. 143 (Table 2a). Wightman's forensic investigations and mapping of Scottish land ownership revealed that, in 2012, 432 landowners possessed half the private rural land in Scotland, some 7.9 million acres.

13 See entries for *dreich, smirr, spindrift* and *drookit* in the Dictionaries of the Scots Language (https://dsl.ac.uk/) and Collins English Dictionary (https://www.collinsdictionary.com/dictionary/english/). For *stoating* (and yet more Scottish words for 'rain'), see '15 Words which Can Only Be Used to Describe Scottish Weather', *The Scotsman*, 19 April 2016, https://www.scotsman.com/whats-on/arts-and-entertainment/15-words-which-can-only-be-used-describe-scottish-weather-1478371.

14 The scientific name for glue crust fungus is *Hymenochaete corrugata*.

15 Both figures from Alliance for Scotland's Rainforest, 'Saving Scotland's Rainforest', https://savingscotlandsrainforest.org.uk/.

16 Alliance for Scotland's Rainforest, 'About Scotland's Rainforest', https://savingscotlandsrainforest.org.uk/rainforest.

17 Wightman, *The Poor Had No Lawyers*, p. 64.

18 John Prebble, *The Highland Clearances*, 1963, p. 24.

19 Wightman, *The Poor Had No Lawyers*, p. 222.

20 John Lister-Kaye, *Ill Fares the Land: a Sustainable Land Ethic for the Sporting Estates of the Highlands and Islands*, 1994, quoted in Wightman, *The Poor Had No Lawyers*, p. 220.

21 Quoted in Prebble, *The Highland Clearances*, p. 203.

22 Wightman, *The Poor Had No Lawyers*, p. 222.

23 Andy Wightman's investigations have shown that, by the early twenty-first century, there were around 340 huge sporting estates in Scotland covering 5 million acres: *The Poor Had No Lawyers*, p. 219.

24 'Around 100,000 after the Second World War': cited by Simon Pepper, 'A Brief History of "the Deer Problem" in Scotland', Forest Policy Group, 15 February 2016, http://www.forestpolicygroup.org/blog/a-brief-history-of-the-deer-problem-in-scotland/. 'As many as 1 million today': Deer Working Group, 'The Management of Wild Deer in Scotland', February 2020, https://www.gov.scot/publications/management-wild-deer-scotland/pages/2/.

25 Pepper, 'A Brief History of "the deer problem"'.

26 Twitter thread by @Collbradan, https://twitter.com/collbradan/status/14531080 88534454276?s=20&t= CbP12yMrScZ0kuV8C4rNfQ. The ecologist Alan Watson Featherstone has tweeted about similar deer browsing pressures at Glen Beasdale in Morar: https://twitter.com/ AlanWatsonFeat1/ status/1496879185691914246.

27 Deer Working Group, 'The Management of Wild Deer in Scotland': 'The requirements of a control scheme make implementing one a power of last resort and there has not been a control scheme in the 60 years since the power was first introduced in 1959.'

28 R. J. Putman, 'Deer Management Plan for the Morvern Deer Management Area, 2015–2020', pp. 94–5, https://morverndmg.deer-management.co.uk/wp-content/uploads/2016/06/MDMG-2015-to-2020-DMP-Complete-small-file.pdf. No more recent plan is yet available.

29 D. Campbell et al., 'Trends in Woodland Deer Abundance across Scotland: 2001–2016', Scottish Natural Heritage Commissioned Report No. 948, 2017, p. ii, https://www.nature.scot/sites/default/files/2017-11/Publication%20 2017%20-%20SNH%20 Commissioned%20Report%20 948%20-%20Trends%20in%20 woodland%20deer%20 abundance%20across%20 Scotland%202001-2016.pdf.

30 Scottish Environment LINK, 'Managing Deer for Climate, Communities and Conservation', 2020, https://s3.documentcloud.org/documents/6617226/Link-Deer-Report.pdf.

31 The Deer Working Group published its final report and set of ninety-nine

recommendations in February 2020: https://www.gov.scot/publications/management-wild-deer-scotland/. The Scottish government published its response in March 2021, accepting ninety-one of these recommendations: https://www.gov.scot/publications/deer-working-group-recommendations-scottish-government-response/. Following the May 2021 Scottish elections, the SNP and Greens reached a power-sharing deal; the shared policy programme agreed to implement the recommendations of the Deer Working Group: https://www.gov.scot/publications/scottish-government-and-scottish-green-party-shared-policy-programme/.

32 Alliance for Scotland's Rainforest, 'About Scotland's Rainforest', https://savingscotlandsrainforest.org.uk/rainforest.

33 Sandy Coppins and Brian Coppins, *Atlantic Hazel: Scotland's Special Woodlands*, 2012, pp. 12 and 8.

34 Billy Fullwood, a young botanist living in Cornwall, has discovered numerous stands of hazel supporting hazel gloves fungus on the outskirts of Bodmin, and in many other locations around Cornwall. Follow him on Twitter: @BotanyCornwall.

35 NatureScot, 'The Beloved Hazel', 30 April 2021, https://scotlandsnature.

blog/2021/04/30/the-beloved-hazel/. See also Hugh Asher, 'The Gaelic Tree Alphabet', Darach Croft, 20 September 2021, https://darachcroft.com/news/the-gaelic-tree-alphabet.

36 Oliver Rackham, *Trees and Woodland in the British Landscape*, 1976, p. 72.

37 Coppins and Coppins, *Atlantic Hazel*, p. 82.

38 Forest Research, 'Managing and Controlling Invasive Rhododendron', 2006, p. 1, https://assets.publishing.service.gov.uk/government/uploads/system/uploads/attachment_data/file/698576/managing_and_controlling_rhododendron.pdf.

39 Richard Milne, *Rhododendron*, 2017, p. 161. See also Katharina Dehnen-Schmutz and Mark Williamson, '*Rhododendron ponticum* in Britain and Ireland: Social, Economic and Ecological Factors in Its Successful Invasion', *Environment and History*, 12:3 (2006), pp. 325–50, https://www.jstor.org/stable/20723582.

40 Milne, *Rhododendron*, p. 164.

41 Ibid., p. 162.

42 Forestry Commission, 'NFI Preliminary Estimates of the Presence and Extent of Rhododendron in British Woodlands', 2016, https://www.forestresearch.gov.uk/documents/2715/Presence_of_Rhododendron_in_British_Woodlands.pdf. The report estimates that rhododendron covers 37,600 hectares of

England, 53,300 hectares of Scotland and 7,900 hectares of Wales: 98,800 hectares in total, or 244,000 acres. Birmingham is around 268 square kilometres (Encyclopedia Britannica: https://www.britannica.com/place/Birmingham-England), or 66,000 acres (× 4 = 264,000 acres).

43 Gordon Gray Stephens and Bob Black, 'Rhododendron in the Rainforest: Approaches to a Growing Problem', Woodland Trust Scotland, September 2021, p. 9, https://www.woodlandtrust.org.uk/publications/2021/10/rhododendron-in-the-rainforest?fbclid=IwAR0vZx1X6iSrEggdMASVTEGPLt8W1gvbTTQ-xF20ro--R38ku7rnIGZHlxc. The report states that 12,000 hectares (30,000 acres) is infested with rhododendron, i.e. 40 per cent of the total.

44 Forestry Commission Scotland (now Scottish Forestry), 'An Approach to Prioritising Control of Rhododendron in Scotland', 2017, https://forestry.gov.scot/publications/26-an-approach-to-prioritising-control-of-rhododendron-in-scotland. See also Scottish Forestry's GIS maps data: https://open-data-scottishforestry.hub.arcgis.com/datasets/ffec009f77d04f63a98ba97205ef74ab_0/explore?location=56.552335%2C-5.523089%2C7.88.

45 For more information on the work that ACT is doing to remove rhododendron from

Islay and other parts of Argyll, see ACT, 'Pesky Ponticum (Part One): What's so Bad about Rhododendron?', 18 March 2019, https://www.act-now.org.uk/blog/pesky-ponticum-part-one-whats-so-bad-about-rhododendron, and The CANN Project, 'Rinns of Islay', https://thecannproject.org/explore/rinns-of-islay/.

46 Lorna Siggins, 'Poor Management of Dead Rhododendrons "Fuelled Killarney National Park Fire"', *The Times*, 17 May 2021, https://www.thetimes.co.uk/article/poor-management-of-dead-rhododendrons-fuelled-killarney-national-park-fire-wh8dlhgr6. See also the Twitter thread by Eoghan Daltun (@IrishRainforest), https://twitter.com/irishrainforest/status/1393809430978994176.

47 Conservationists I've spoken to involved in rhododendron eradication recommend stem injections with glyphosate.

48 Alliance for Scotland's Rainforest, 'Saving Morvern's Rainforest', https://savingscotlandsrainforest.org.uk/asr-projects/saving-morverns-rainforest.

49 Scottish Environment LINK, 'Case Study 14: Saving Morvern's Rainforest – Alliance for Scotland's Rainforest', https://www.scotlink.org/wp-content/uploads/2021/02/Case-Study-14.pdf. See also Martin Williams, 'Cash Crisis: Clouds Gather in the Mission to Save Scotland's Rare

Rainforests', *The Herald*, 27 January 2021, https://www.heraldscotland.com/news/19040862.cash-crisis-clouds-gather-mission-save-scotlands-rare-rainforests/.

50 The Scottish government funds rhododendron removal via its Forestry Grant Scheme under the Scottish Rural Development Plan. Similar schemes exist in England and Wales. For example, see Scottish Forestry, 'Invasive Rhododendron', https://forestry.gov.scot/forests-environment/biodiversity/non-native-species/invasive-rhododendron.

51 Richard Baynes, 'Killer Shrub Will Cost £400m to Eliminate, Say Conservationists', *The Times*, 4 June 2018, https://www.thetimes.co.uk/article/killer-shrub-will-cost-400m-to-eliminate-say-conservationists-6tgm62nqq.

52 £10 million estimate: Snowdonia National Park Authority, 'Rhododendron in Snowdonia and a Strategy for Its Control', 2008, 'Executive Summary', https://www.snowdonia-npa.gov.uk/__data/assets/pdf_file/0031/164785/Rhododendron-Strategy-Final.pdf. £45 million estimate: R. H. Gritten, '*Rhododendron ponticum* and Some Other Invasive Plants in the Snowdonia National Park', in P. Pyšek et al. (eds), *Plant Invasions: General Aspects and Special Problems*, 1995, pp. 213–19.

53 Robert Burns, 'Verses on the Destruction of the Woods near Drumlanrig', 1791.

54 See http://www.whoownsscotland.org.uk/.

55 Community Land Scotland states that 560,000 acres of Scotland is currently owned by community groups: 'About Community Land Scotland', https://www.communitylandscotland.org.uk/.

56 For more about Friends of Glenan Wood, and to make a donation, visit https://www.glenanwood.org.uk/.

57 See https://www.cormonachan-woodlands.co.uk/.

58 See https://kilfinancommunityforest.co.uk/. For more on woodland crofts, see the Community Woodlands Association, 'Woodland Crofts', https://www.communitywoods.org/woodland-crofts.

59 Thanks to Matilda Scharsach at Woodland Trust Scotland for corresponding with me about this brilliant project.

60 Author's conversation with Gordon Gray Stephens. See also ACT's report on their work in removing rhodies in Glen Creran: https://www.act-now.org.uk/glen-creran.

61 Gray Stephens and Black, 'Rhododendron in the Rainforest', p. 28.

62 See, for example: James Fair, 'Why Has a Commercial Rewilding Firm Been Accused of Eco Colonialism?', *ENDS Report*, 4 November 2021, https://www.endsreport.com/article/1732347/why-commercial-rewilding-firm-

accused-eco-colonialism; Jon
Hollingdale, 'Green Finance,
Land Reform and a Just
Transition to Net Zero: A
Discussion Paper', Community
Land Scotland, February 2022,
https://www.communityland
scotland.org.uk/wp-content/
uploads/2022/03/Green-
finance-land-reform-and-a-just-
transition-to-net-zero-a-
discussion-paper_CLS_
Feb-2022.pdf; and Magnus
Davidson, 'Land, Green Lairds
and Rewilding', *Bella
Caledonia*, 20 February 2022,
https://bellacaledonia.org.
uk/2022/02/20/land-green-
lairds-and-rewilding/.

63 See https://twitter.com/
DailyGael/status/
1448192514712903680 and
responses.

64 John T. Koch, *Celtic Culture: A
Historical Encyclopedia*, 2006, p.
775: 'Gaelic . . . the original
sense of the ethnonym is,
therefore, "forest people", hence
"wild men, savages".'

10. Bringing Back Britain's Rainforests

1 NBN Atlas, '*Schistostega
pennata*', https://species.
nbnatlas.org/species/
NHMSYS0000310606.

2 Cabilla Cornwall, 'Our Team',
https://www.cabillacornwall.
com/our-vision/about/.

3 Victoria Gill, 'UK Public Now
Eating Significantly Less Meat',
BBC News, 8 October 2021,
https://www.bbc.co.uk/news/
science-environment-58831636.

4 For more on the science behind
nature contact and mental and
physical health, see Lucy Jones,
*Losing Eden: Why Our Minds
Need the Wild*, 2020.

5 Rewilding Britain, 'Cabilla
Cornwall', https://www.
rewildingbritain.org.uk/
rewilding-projects/cabilla-
cornwall.

6 E. O. Wilson Foundation, 'E.
O. Wilson Interviewed on
BBC's "The Life Scientific"', 27
September 2015, https://
eowilsonfoundation.
org/e-o-wilson-interviewed-on-
bbcs-the-life-scientific/.

7 Natural History Museum,
'Natural History Museum
Reveals the World Has Crashed
through the "Safe Limit for
Humanity" for Biodiversity
Loss', 11 October 2021,
https://www.nhm.ac.uk/press-
office/press-releases/
natural-history-museum-reveals-
the-world-has-crashed-
through-the.html. See also the
Natural History Museum's
'Biodiversity Trends Explorer':
https://www.nhm.ac.uk/
our-science/data/biodiversity-
indicators.html.

8 DEFRA, 'Wild Bird
Populations in the UK, 1970–
2019', 9 December 2021, p. 5,
https://assets.publishing.service.
gov.uk/government/uploads/
system/uploads/attachment_
data/file/1039187/UK_
Wild_birds_1970-2020_
FINAL.pdf, which shows a 57
per cent decline in farmland
bird species since 1970.
DEFRA, 'Butterflies in the
United Kingdom: Habitat

Specialists and Species of the Wider Countryside, 1976 to 2020', September 2021, p. 13, https://assets.publishing.service. gov.uk/government/uploads/ system/uploads/attachment_ data/file/1018259/ Butterflies_in_the_UK_1976_ to_2020_final_v0.3_accessible. pdf: 'The wider countryside woodland butterfly index for the UK decreased by 42% between 1990 and 2020'. Rothamsted Research, 'The State of Britain's Larger Moths', https://insectsurvey.com/moth- state: 'The total abundance of moths declined by 28% over the period 1968–2007.' Other recent studies from other countries have indicated even steeper declines in insect abundance in recent decades.

9 Around 500 species of lichen: Plantlife, 'Lichens and Bryophytes of Atlantic Woodland in Scotland: An Introduction to Their Ecology and Management', 2010, p. 6, https://www.plantlife.org.uk/ application/files/9914/8233/ 4028/PLINKS_Atlantic Woodland_LRes.pdf. Over 160 species of mosses and liverworts: Derek Ratcliffe, 'An Ecological Account of Atlantic Bryophytes in the British Isles', *New Phytologist*, 67 (1968), pp. 365–439.

10 Dominick DellaSala (ed.), *Temperate and Boreal Rainforests of the World: Ecology and Conservation*, 2011, pp. 25 and 28–29.

11 Neil Sanderson et al.,

'Guidelines for the Selection of Biological SSSIs, Part 2: Detailed Guidelines for Habitats and Species Groups – Chapter 13: Lichens and Associated Microfungi', Joint Nature Coordinating Committee (JNCC), 2018, p. 5, https://data.jncc.gov.uk/ data/330efebf-9504-4074-b94c- 97e9bbdbe746/SSSI- Guidelines-13-Lichens-2018. pdf.

12 According to the ecologist Christopher Ellis, 'The British Isles have the most strongly oceanic of European climates . . . and therefore account for c.40% of bioclimatic space suitable to the development of European rainforest': 'Oceanic and temperate rainforest climates and their epiphyte indicators in Britain', *Ecological Indicators*, 70 (2016), pp. 125–33, https://www. researchgate.net/ publication/304105743_ Oceanic_and_temperate_ rainforest_climates_and_ their_epiphyte_indicators_in_ Britain.

13 Henry Dimbleby et al., *National Food Strategy: The Plan*, 2021, pp. 91–3, and *National Food Strategy: The Evidence*, 2021, p. 41, at https://www. nationalfoodstrategy.org/ the-report/.

14 'PM Boris Johnson's Address to the COP26 Forests & Land-Use Event', 2 November 2021, https://www.gov.uk/ government/speeches/pm-boris-

johnsons-address-to-the-cop26-forests-land-use-event.

15 Neil Sanderson, 'Lichen Survey of Millook Valley Woods, Cornwall', unpublished report for the Woodland Trust, February 2011. Thanks to Malcolm Allen at the Woodland Trust for sharing the report, and for showing me around Millook Woods.

16 Natural England, 'Condition of SSSI Units for Site Johnny Wood SSSI', current as of 27 May 2022, https://designatedsites.naturalengland.org.uk/ReportUnitCondition.aspx?SiteCode=S1000280&ReportTitle=Johnny%20Wood%20SSSI.

17 George Peterken, 'Reversing the Habitat Fragmentation of British Woodlands', WWF-UK report, January 2002, p. 5, https://wwfeu.awsassets.panda.org/downloads/ukforestsfragmentation.pdf.

18 Richard K. Broughton et al., 'Long-Term Woodland Restoration on Lowland Farmland through Passive Rewilding', *PLoS ONE*, 16:6, e0252466, https://doi.org/10.1371/journal.pone.0252466: 'Woody vegetation colonisation of the New Wilderness was rapid, with 86% vegetation cover averaging 2.9 m tall after 23 years post-abandonment.' See also Figure 1, showing regeneration of trees at the 'New Wilderness' site out to distances of 100 to 150 metres from the nearest seed source.

19 GIS mapping analysis by Tim Richards and Guy Shrubsole, 2022. Taking our maps of ancient woodland in Britain's oceanic zone, Tim drew a 150-metre wide 'buffer zone' around each wood. When applying this to England's rainforests, Tim was also able to factor in a set of constraints he first developed for a Friends of the Earth project to map opportunities for woodland creation in England. These included allowing no trees on peat soils, on other priority habitats like species-rich grassland, or on high-quality farmland (Agricultural Land Classification grades 1–3a). There are around 46,000 acres of existing temperate rainforest in England; allowing these sites to spread outwards by 150 metres, whilst factoring in the aforementioned constraints, adds around 37,000 acres of regenerated rainforest – an 80 per cent increase, almost doubling the rainforest area. Across Britain as a whole, we think there are around 333,000 acres of surviving rainforest. Allowing this to expand outwards by 150 metres on all sides results in a *quadrupling* of rainforest area, but this does not take into account the environmental constraints we were able to apply in England. We are confident, however, that a near-doubling of rainforest cover across Britain would be possible once those constraints were applied.

20 Miles King, 'Berrier Farm under Trees: 100 Acres of Peat Bog, Heat and Wildlife-Rich Grassland Destroyed by Tree Planting', A New Nature Blog, 20 November 2020, https://anewnatureblog.com/2020/11/06/berrier-farm-under-trees-100-acres-of-peat-bog-heath-and-wildlife-rich-grassland-destroyed-by-tree-planting/.

21 To see the bracken maps and for more details on how bracken can guide rainforest restoration, see my 'Using Bracken as a Guide for Regenerating Rainforest', Lost Rainforests of Britain, 3 February 2022, https://lostrainforestsofbritain.org/2022/02/03/using-bracken-maps-as-a-guide-for-regenerating-rainforest/.

22 For more on Sir Richard Acland and his political ideals, see the National Trust, 'The Acland Family', https://www.nationaltrust.org.uk/killerton/features/the-acland-family. For more on Killerton, Horner Wood and the surrounding Holnicote Estate on Exmoor, see the National Trusts, 'Killerton House', https://www.nationaltrust.org.uk/killerton/features/killerton-house, and 'Horner Wood', https://www.nationaltrust.org.uk/horner-wood.

23 For my map of who owns Dartmoor, see my 'Who Owns Dartmoor?', 13 March 2021, Who Owns England?, https://whoownsengland.org/2021/03/22/who-owns-dartmoor/.

24 Wild Card, 'Open Letter: It's Time to Rewild Royal Land', 9 June 2021, https://www.wildcard.land/letter/.

25 For more information about farmer clusters, see https://www.farmerclusters.com/ and, for the Dartmoor cluster, https://www.farmerclusters.com/profiles/south-west/csff120008-dartmoor/.

26 J. E. G. Good et al., 'Developing New Native Woodland in the English Uplands', English Nature Research Report 230, February 1997, pp. 43 and 76 http://publications.naturalengland.org.uk/publication/62057?category=47017. In the section of the report which considered woodland creation on Dartmoor in detail, the authors asked in reference to Wistman's Wood: 'Should we not be considering, at least, the establishment of similar types of high altitude oakwoods in appropriate locations on Dartmoor?' (p. 42). Though they felt there were more opportunities for woodland creation in Dartmoor's eastern valleys, the authors also concluded: 'That is not to say that no consideration should be given to the possibility of creating new woods to mimic (as far as that is possible) the high altitude woods such as Wistman's wood; it would certainly present a challenge!' (p. 43).

27 'Conserve and Enhance Natural Beauty', Dartmoor National Park Partnership Plan 2021–2026, https://www.yourdartmoor.org/the-plan/better-for-nature/conserve-and-enhance-natural-beauty, features a pledge of 2,000 hectares (5,000 acres).

28 Moor Trees, 'A New Vision for Dartmoor', https://moortrees.org/about-us/vision-for-dartmoor/

29 See the Alliance for Scotland's Rainforests website: https://savingscotlandsrainforest.org.uk/.

30 See the Celtic Rainforests Wales website: https://celticrainforests.wales/.

31 Eoghan Daltun, 'What Is Beara Rainforest All About?', Beara Rainforest, http://beararainforest.com/. Follow Eoghan on Twitter: @IrishRainforest.

32 Hugh Wilson in Jordan Osmond and Antoinette Wilson, *Fools and Dreamers: Regenerating a Native Forest*, Happen Films, 2019, https://www.youtube.com/watch?v=3VZSJKbzyMc. 'Banks Peninsula appears to fall into the cool temperate, oceanic, subhumid bioclimatic cell,' stated Wilson in his assessment of the area's climate and vegetation prior to the regeneration of the rainforest: Hugh D. Wilson, 'Bioclimatic Zones and Banks Peninsula', *Canterbury Botanical Society Journal*, 27 (1993), p. 28, https://bts.nzpcn.org.nz/site/

assets/files/22426/cant_1993_27__22-29.pdf. 'The rigorous removal of browsing pressure is likely to be a key factor promoting rapid establishment of native forest,' Wilson wrote in 1994, and so it has proven: Hugh D. Wilson, 'Regeneration of Native Forest on Hinewai Reserve, Banks Peninsula', *New Zealand Journal of Botany*, 32:3 (1994), p. 381, https://www.tandfonline.com/doi/pdf/10.1080/0028825X.1994.10410480.

33 DellaSala (ed.), *Temperate and Boreal Rainforests*, p. 251. See also Table 10–1, pp. 247–50, which lists the areas of temperate and boreal rainforest strictly protected by each nation. The entry for Great Britain reads simply: 'None'.

34 DEFRA, 'Request for Proposals for the Big Nature Impact Fund Manager', 16 November 2021, p. 6, https://www.contractsfinder.service.gov.uk/Notice/Attachment/01630643-1a08-4e42-ad07-4e1de4e2fb50.

35 Emma Gatten, 'Britain's Lost Rainforests Could Return in Post-Brexit Plans', *Telegraph*, 25 December 2021, https://www.telegraph.co.uk/environment/2021/12/25/britains-lost-rainforests-could-return-post-brexit-plans/.

36 Parliamentary question 110592 tabled by Anthony Mangnall MP on 24 January and answered 8 February 2022,

https://questions-statements.
parliament.uk/written-
questions/detail/2022-01-24/
110592; and parliamentary
question 110593, tabled 24
January and answered 1
February 2022, https://
questions-statements.parliament.
uk/written-questions/
detail/2022-01-24/110593.

11. Ghost Hunting
1 William Crossing, *Guide to
Dartmoor*, 1909, pp. 175–6.
2 T. S. Eliot, *The Waste Land*,
1922, lines 19–20.

INDEX

Index